MW00762752

The Advanced Communications Technology Satellite

We dedicate this book to the many people who formed the ACTS Team whose dedication, cooperation, and expertise made it a successful project.

THE ADVANCED COMMUNICATIONS TECHNOLOGY SATELLITE

An Insider's Account of the Emergence of Interactive Broadband Services in Space

Richard T. Gedney
Ronald Schertler
Frank Gargione

SciTECH
PUBLISHING, INC.

Published by SciTech Publishing, Inc., Mendham, NJ
©2000 by Advanced Communications Technology (ACT) Co.
All rights reserved. No part of this book may be reproduced or used in any form whatsoever without written permission except in the case of brief quotations embodied in critical articles and reviews. For information, contact SciTech Publishing.
Printed in Korea
10 9 8 7 6 5 4 3 2 1

ISBN 1-891121-11-1

SciTech President: Dudley R. Kay
Production and Design Services: *TIPS* Technical Publishing
Cover Design: Janice Bakker and Tom Miller

Chapter 8 opener illustration of iSKY Ka spot-beam satellite courtesy of iSKY. Chapters 1 and 6 opener illustrations courtesy of Via Satellite.

This book is available at special quantity discounts to use as premiums and sales promotions, or for use in corporate training programs. For more information and quotes, please contact:

Director of Special Sales, SciTech Publishing, Inc.
89 Dean Road Mendham, NJ 07095
Phone: (973) 543-1115 Fax: (973) 543-2770
Email: sales@scitechpub.com http://www.scitechpub.com

Information contained in this work has been obtained from sources believed to be reliable. However, neither SciTech Publishing nor its authors guarantee the accuracy or completeness of any information published herein, and neither SciTech Publishing nor its authors shall be responsible for any errors, omissions, or damages arising out of use of this information. This work is published with the understanding that SciTech Publishing and its authors are supplying information but are not attempting to render engineering or other professional services. If such services are required, the assistance of an appropriate professional should be sought.

CONTENTS

FOREWORD

The Advanced Communications Technology Satellite (ACTS) has been an unqualified success in spite of the best efforts of politicians and bureaucrats to kill it. It has outlasted its critics inside and outside of NASA, the Administration, and the Congress.

The fact that the ACTS satellite and its earth stations have met or exceeded program objectives—in a period when even conventional satellites have been plagued by a series of failures—is significant. It is a testimony to careful nurturing of the project by the integrated team of government and industry professionals that labored on despite every annual budget crisis. Their care in design and testing has been proven in orbit—the ultimate test.

The success story of ACTS is being proven in the flood of new commercial applications for over 700 new Ka-band satellites. In reality, if only a fraction of these new satellites are built, the $439 million invested by the U.S. taxpayers on the ACTS satellite will quickly be repaid from the resulting taxes on spacecraft designed, built, and launched by U.S. manufacturers. Even more tax revenue will come from the associated service sector and the sale of two-way earth stations to the public. Few other government programs can demonstrate such a large and rapid return on taxpayer investment. Based on this financial success, NASA should consider an ACTS follow-on program using a different frequency and even newer technology.

The repayment of the ACTS investment is already evident. At my elbow lies a mobile satellite system (MSS) phone (which uses spot beams, onboard processing, and Ka-band feeder links). Ongoing work is being done for various Ka-band applications in several orbits. The movable beam technology is an important element of a new class of satellites currently being developed. These things take time and are nearly impossible to complete without a pathfinder (like ACTS) to showcase the new technology through long-term, in-orbit demonstrations.

This book is unique in that it is a case study of the entire program—including the management, design, operation, and application—of this new technology. It provides useful peeks into the future of the commercial programs it is spawning, and should be a must-read for the managers of those programs. It is also frank about some of the problems awaiting other Ka-band satellites and earth stations, such as the effects of the sun and rain on satellite and earth station antennas. Paying attention to the wisdom contained in this book will save much grief (and perhaps even someone's job).

Walter L. Morgan
Communications Center
2723 Green Valley Road
Clarksburg, MD 20871
United States of America

PREFACE

Technology has traditionally been supported by governments and most often for the purpose of waging war. Fortunately, once a war technology matures, it may be put to peaceful use and find its way into everyday life with immeasurable benefits for all of mankind. Technology development by government has never been simple or lacking in controversy, even for those technologies aimed directly at the betterment of mankind. This book covers one such program, hoping to provide a glimpse of the effort it takes to bring even peaceful technology development to fruition. Active federal participation in the development—as well as the encouragement—of technological innovation has existed for a long time in the U.S. In 1836, for example, Congress appropriated $30,000 to subsidize Samuel Morse's first telegraph—an experimental line from Washington, D.C. to Baltimore.

Government programs have differed in the extent to which they blend three conceptually distinct purposes: contributing basic research from which new technologies will be born, advancing technology for which government is the customer, and developing technology for use by the private sector.

There have been many successes and failures in all three categories, but perhaps the third category is the most difficult for government to sponsor. This is because the success of a program for commercial benefit depends on acceptance in the marketplace. It is difficult enough for the private sector to predict useful technology for future, uncertain markets, but it is even more difficult for public decision-makers who are usually less skilled at forecasting commercial markets than their private sector counterparts. Secondly, the government support of research and development (R&D) for commercial application involves industrial policy, which is normally highly controversial. A key issue in the debate is whether or not the government is likely to do more harm than good in sponsoring technology development.

The Advanced Communications Technology Satellite (ACTS) program is a government program of the third category, and this book is a case study of that type of sponsored research. Cohen and Noll [1] have aptly defined the characteristics of this category of government sponsorship as follows:

- When the project is adopted, commercialization is a sufficiently important objective that the program cannot be considered a success unless the private sector adopts the technology at the conclusion of the program.

- The program is financed by direct federal expenditures through the normal budgetary process with review by the Office of Management and Budget and appropriations by Congress.

- Government officials are deeply involved in technical design and management decisions.

- From the beginning, the government is committed to support the development of prototypes and the first commercial demonstration, not just the early research or a component of a broader private research effort.

- Although the advocates of the programs may have complex reasons for favoring them, each program is justified initially on strictly economic criteria. That is, it will reduce costs or enhance performance by a sufficient amount to make it an economically attractive investment, even though the private sector alone is unwilling or unable to undertake the project.

ACTS was strongly debated over a ten-year period from 1983 to 1993. Both its advocates and its adversaries spent a great amount of time and energy expressing their opinions and taking action to influence the program as it was progressing. For five years, the administration adversaries succeeded in keeping ACTS out of NASA's budget request to Congress, and congressional advocates restored the ACTS budget each time. The difference in philosophy and political rhetoric was as great as for any other federal program.

Our purpose in writing this book is to shed some light on the core policy issues, the difficulties in executing programs that deal with continuous public policy disagreement, the usefulness of the developed technology to the private sector, and the lessons learned. The authors were all closely associated with the program and have to be considered "advocates." We have made a deliberate attempt, however, to present both sides of the debate and to make the material as factual as we can. ACTS is also a story about the emergence of interactive, broadband services in space. During 1980, when the ACTS program was being formulated, interactive broadband services with the capability for bandwidth-on-demand were identified as the most important new services on the horizon. A few visionary people (the Internet didn't exist and the PC was in its infancy) knew that lease lines for distributing data and video would be too costly. As a result, the central requirements for ACTS included the capability for bandwidth-on-demand from the very beginning. Recognizing the need for transparency, ACTS was also designed as a combined satellite/terrestrial network with seamless interconnectivity. As a result, when ACTS was launched 13 years later it was ready to show the way into the information age. The user trials that were conducted using ACTS demonstrated how valuable this new capability could be, and proved that satellites would be an integral part of the twenty-first century's information infrastructure.

Overview of the Book

This book is written for the managers of aerospace engineering projects, public and private policy makers, technology planners and developers, and the satellite communications community. The first chapter of the book gives a history of satellite communications prior to ACTS, so the reader has an understanding of the development of the communications satellite industry and the role that government played up to the introduction of ACTS. Chapter 1 also discusses the formulation of the program, its objectives, a brief description of the technology, the advocate and adversary viewpoints, and the contractual process. Chapters 2 and 3 provide a detailed description of satellite and earth station technology with a technical assessment of the technology based on in-orbit results. Chapter 4 presents the applicability of the technology based on the results of user trials. Chapter 5 outlines the ACTS Ka-band propagation campaign, provides quantitative 30/20 GHz rain fade propagation measurements, and presents some methods for mitigating the large fades.

Because of the continual controversy surrounding ACTS, program execution was severely impacted. The difficulties imposed by this controversy, as well as those due to technical and managerial problems, are discussed in Chapter 6. Also presented in Chapter 6 is a discussion of the practices that were used to ensure that ACTS was a 100% technical success.

Extensive market research was conducted during formulation of the ACTS program to identify future satellite markets and their demand. It is always difficult to conduct accurate market research. In Chapter 7, the results of market research that was used in program formulation are discussed as well as the accuracy of their predictions. As this chapter points out, the market research was quite visionary and played a key role in identifying the necessary technology.

Chapter 8 makes an assessment of how well the ACTS technology is being utilized by the private sector. Finally, Chapter 9 takes a look at the recent impact of the ACTS program on commercial communications and contrasts this with the impact of military as well as commercial R&D to gain a perspective on the need for government sponsorship of technology. The chapter then presents the major arguments for and against the government sponsorship of a flight program to demonstrate and prove the high-risk technology. The two arguments for the ACTS flight program were: 1) the need to counterbalance foreign government subsidies to maintain U.S. preeminence, and 2) the service industry's perception that implementing ACTS technology without a flight demonstration was too great a technical risk. The arguments against the flight program were: 1) that it created an unfair competitive advantage for the winning contractors, and 2) that the government was not capable of successfully guiding a technology program for commercial application. Chapter 9 ends with a discussion of what the role of government technology sponsorship should be in the future.

This book covers a very broad range of technical, managerial, and government policy issues. For those who are non-technical, Chapters 2-5 can be skimmed and the reader will still obtain a good understanding of the management and government policy aspects of ACTS.

The opinions expressed in this book are those of the authors and do not represent the views of any of the government organizations or companies that participated in the ACTS program. A large number of very dedicated companies, organizations, and people worked on ACTS, and that entire team is responsible for its success. Because of the limited space available, we deeply regret that it is not possible to include the names of all the people and organizations that participated in the program and who made very significant contributions.

It is certain that government sponsorship of technology for commercial application will continue. The government's program for the Next Generation Internet is a good example of such continuing sponsorship. We hope, therefore, that this book aids in formulation of improved public policy regarding government sponsorship of technology.

Acknowledgments

In researching and assembling materials for this book, we have relied on the assistance of a great number of people, most of whom are acknowledged in the pages that follow. We would especially like to thank the NASA Glenn Research Center in Cleveland, Ohio, for its assistance in making much of material available and in encouraging its development. In particular we wish to thank from NASA Glenn Roberto Acosta, Robert Bauer, Richard Krawczyk, Rodney M. Knight, who recently retired, Louis Ignaczak, Pete Vrotsos, and Mike Zernic.

Joanne Poe of Flowen LLC in Cleveland was very helpful in providing information on the role of government in sponsoring technology in the future.

Finally, we wish to thank Dudley Kay of SciTech Publishing, Inc. of Mendham, New Jersey, who published the book, and supported the project from the time he first saw the manuscript, and Robert Kern, publisher, and Lynanne Fowle, managing editor, at TIPS Technical Publishing in Carrboro, North Carolina for producing this book on a very ambitious schedule.

CHAPTER 1

PROGRAM FORMULATION

History has shown that leadership in communication carries with it substantial economic and political advantages. Communication is essential in all aspects of life, including business, national affairs, and cultural well-being. Governments have always been involved in the regulation and technology of communication. In the United States, the Federal Communications Commission (FCC) is the primary agency for communication regulation, and the National Aeronautics and Space Administration (NASA) and Department of Defense (DOD) programs have been responsible for many space communication technology innovations since the late 1950s.

One topic covered in this book is the proper role of government in sponsoring space communications research. The difficulty in the government's sponsoring of space communications research is that it benefits some companies while putting others at a perceived or real competitive disadvantage. When NASA or DOD develops *mission-specific* communication technology, much of it becomes very useful in commercial satellite communication. This type of technology development, however, is viewed by the industry as nonintrusive and acceptable.

The Advanced Communications Technology Satellite (ACTS) program had a different objective. Its prime purpose was to develop technology for use in providing *commercial* communication (not a government-specific mission), and it became embroiled in a continuous controversy over the proper role of government for this purpose. The primary argument in favor of the program was that it was important for the government to ensure the U.S. preeminence in satellite communication and develop the technology for efficient use of the frequency spectrum. The main objections to NASA's sponsorship of the ACTS program was that it created an unfair competitive advantage for some companies and that it was not possible for the government to predict what technology was needed.

The next section reviews the early development of satellite communication so the reader has an understanding of the significant technological role that the federal government played prior to ACTS. At the start of the space era, relatively little was known about the harsh environment of space and the life expectancy of a satellite. Research and development carried out by NASA and DOD in the early years of the space program served to reduce the technological risks associated with the launch and operation of communication satellites. NASA's and DOD's sponsorship of satellite communication technology—starting after Sputnik—helped to foster the rapid growth of the now burgeoning global satellite communication industry, which is projected to grow from the $51.9 billion spent in 1998 to $114.9 billion in 2003 [2].

The Early History of Satellite Communication

Communication satellites were first flown and tested in the early 1960s. Following the opening of the space age by the Soviet Union's Sputnik satellite in 1957, NASA and DOD became prime movers in a series of communication satellite projects [3,4,5,6]. Those early, pioneering efforts in the development of satellite communication technology are listed here along with their accomplishments.

Echo Balloons The use of a large aluminum-covered sphere as a passive reflector in orbit was established with Echo I in 1960 and Echo II in 1964. Although the NASA-sponsored Echoes never proved practical, the development of earth stations to operate with them proved useful in subsequent projects with active repeater satellites carrying onboard receivers and transmitters.

Echo Balloon undergoing tests prior to launch.

Telstar In 1962, two active communication satellites (Telstar I & II), designed and built by AT&T's Bell Telephone Laboratories, were placed into elliptical orbits. Telstar I was the world's first active repeater satellite when launched on

3

July 10, 1962. Telstar flights demonstrated the transmission of two-way telephone, TV, data, and facsimile via satellite, which subsequently became standard services. It is important to note that AT&T paid for both the satellites *and* their launches.

Syncom The Syncom program was intended to demonstrate geosynchronous communications [7]. The environment of the geosynchronous orbit was unknown, and there was concern that a satellite would not survive long enough to be useful. Project Syncom demonstrated the feasibility of placing a satellite in geosynchronous orbit and maintaining precise stationkeeping and orbit control. Hughes Aircraft Company built three satellites (Syncom I, II, and III) under a *sole-source*, $4 M NASA-U.S. Air Force contract. Syncom II was the first successful geosynchronous satellite. The 86-pound satellite was launched on July 26, 1963. Stationed over the Atlantic Ocean, Syncom II transmitted the first TV signals from space. The phrase "live by satellite" quickly became a buzzword in the world's vocabulary. Syncom III, launched on August 19, 1964, was placed in orbit over the Pacific Ocean. Syncom III was launched just in time to relay TV signals from the 1964 Olympics in Tokyo. The Syncom program, undertaken with the full knowledge that many respected authorities in the communication industry were opposed to geostationary communication satellites, is a good example of a high-risk government program that resulted in large benefits for commercial industry.

Application Technology Satellites NASA's role in satellite communication remained in high gear through the 1960s and the early 1970s. The Application Technology Satellite (ATS) series was initiated to develop new technology, techniques, and services for future communication systems. Six spacecraft were flown between 1966 and 1974. The first five were sole-sourced to Hughes Aircraft in 1964. The government changed its philosophy in 1970, however, and used a competitive procurement process for the sixth spacecraft. The ATS program demonstrated technologies such as *despun* antennas and the use of multiple frequencies. Of equal importance was its role as a platform for demonstrating new services such as direct satellite TV broadcast, telemedicine, tele-education, disaster communication, and mobile communication. The ATS program also showed that a platform in geostationary orbit could be used for weather forecasting and remote sensing by observing almost an entire hemisphere of our planet at once.

ATS 1 through ATS 5 helped to foster the development of a whole series of commercial communication satellites beginning with the Intelsat 3 series and including Westar, COMSTAR, Telstar, and Galaxy. ATS 6 influenced the use of the L-band frequency for the Inmarsat series of maritime communication satellites and fostered a whole series of 3-axis-stabilized commercial satellites including SATCOM, Insat, Arabsat, and TDRSS.

Communications Technology Satellite Another important satellite communication project in the mid-1970s was the Communication Technology Satellite (CTS) [8]. CTS was a joint effort between NASA and the Canadian Department of Communication. The primary purpose of this program was to demonstrate the viability of the 14/12 GHz (Ku) band for satellite communication and to experiment with broadcasting television signals from a high-powered satellite to small, low-cost user terminals. Launched on January 17, 1976, CTS operated successfully into 1979. Its high-powered transmitter allowed the use of small, less-expensive ground terminals. A portable earth terminal—less than 1 meter in aperture size—made the satellite easily accessible for a variety of uses. Following NASA's efforts with CTS, a number of commercial communication satellites incorporated Ku-band frequencies.

International Communication Satellites

After the successful demonstration of Syncom II, Congress created the privately owned Communications Satellite Corporation (COMSAT). COMSAT was given the mandate embodied in the Communications Act of 1962—to establish a global commercial communication system in cooperation with other countries, via the newly formed International Telecommunication Satellite (INTELSAT) organization. INTELSAT ushered in the era of international satellite communication with "Early Bird" (later renamed INTELSAT 1). INTELSAT 1, a commercial Syncom follow-on, was built by Hughes Aircraft and launched on April 6, 1965. The INTELSAT spacecraft bus closely modeled the Syncom III, but the communication subsystem used C-band frequencies pioneered by Telstar. INTELSAT revolutionized transoceanic communications. There was rapid transition from the experimental Syncoms to the operational use of satellites with INTELSAT 1 for transoceanic telephone and television service. In 1969, just one week after a global network of three INTELSAT satellites had been deployed over the Atlantic, Indian, and Pacific Oceans, a record 500 million people watched the first Apollo moon landing live, via satellite. To date, eight generations of INTELSAT have been deployed and a ninth is currently being developed. As of early January 2000, 63 INTELSAT satellites have been launched. No amount of undersea cable circuits could have matched the performance of satellites in that time period, or provided enough circuits to satisfy the demand.

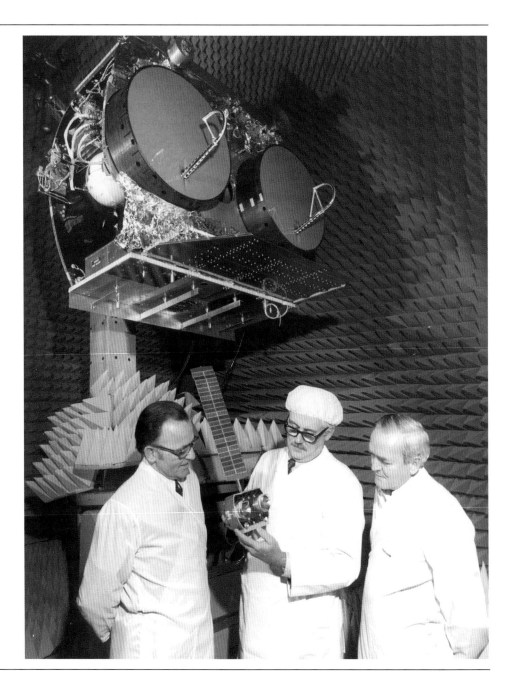

Engineers examine a model of the spacecraft in the anechoic chamber where the CTS payload is undergoing RF tests.

Domestic Communication Satellites

Plans for using satellites to provide domestic communication within the United States created a myriad of political and regulatory dilemmas. In 1965, the American Broadcasting Company (ABC)—in collaboration with Hughes

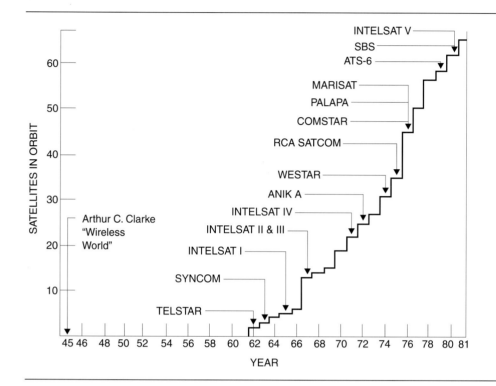

History of Communication Satellites

Aircraft—applied to the Federal Communications Commission (FCC) for permission to create a satellite system to distribute network television programs to their affiliates throughout the Unites States. The FCC returned the request to ABC without acting on it after COMSAT filed in opposition. What followed was a prolonged struggle involving the FCC the White House, and the communications industry.

Meanwhile, Telesat Canada took the initiative and launched Anik 1, the first domestic communication satellite in North America. Launched on November 10, 1972, (seven years after INTELSAT 1), Anik provided television, radio, and telephone services to all but the most northern reaches of Canada. Its shaped antenna beam—a first—concentrated power within a primary-receiving contoured area. In 1973, RCA inaugurated satellite communication service in the United States by using the Anik satellite.

The advent of the Nixon administration brought forth new faces and a change in policy. In 1972, the U.S. domestic "Open Skies" policy became a reality. This policy allowed any qualified carrier to set up a satellite service. Orbital positions (slots) were to be assigned on a first-come, first-served basis. The marketplace would determine the pace of communication satellite development. U.S. industry was now free to provide domestic satellite services.

7

The first U.S. domestic communication satellite was Western Union's Westar 1, which was launched on April 13, 1974. This satellite included 12 transponders, each capable of transmitting either a color television channel, 1,000 telephone circuits, or 60 megabits of computer data. Westar 1 was followed by RCA's SATCOM 1 on December 12, 1975. SATCOM 1 was the first satellite used for cable television. It also provided the first dual-polarized, 24-channel transponder payload, as well as three-axis stabilization. In the next five years, commercial communication satellites proliferated over North America. By 1979, twelve communication satellites, built by several different companies, were in orbit and providing commercial services. Satellite communications became a viable industry, with companies actively competing for market share. All of these satellite systems exploited technology developed by NASA and DOD from 1960 to 1973.

NASA Communication Research Phase-Down

In 1973, the NASA satellite communication program was sharply curtailed. From a 1973 peak of $170M, NASA's communication program funding fell to about $20M by 1975 [9]. This phase-down was the result of budget pressures and the belief that the private U.S. satellite industry was sufficiently mature and healthy to assume the long-range, high-risk technology developments that had been initially sponsored by NASA.

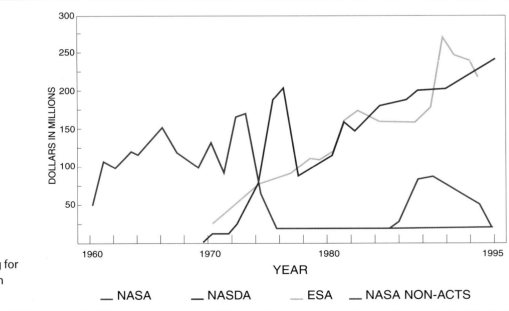

Annual Funding for Communication Programs

Given the existence of profitable and growing satellite manufacturers and service providers, some regarded any federal funding of research and development for the commercial satellite communication industry as unwise. Government programs, they argued, would compete with private sector programs—favoring one element of the private sector over another or encouraging over- or under-investment. The proponents of this argument expected industry, which is motivated by the desire for profit, to undertake the necessary R&D on its own.

The NASA ATS 6 and CTS projects were well underway in 1973. These were allowed to finish, but the ATS G and ATS H/I spacecraft initiatives, not yet underway, were cancelled. After this curtailment, NASA only maintained a minimal involvement in advanced technology development for commercial communication satellites and initiated no new flight programs.

From this point in time, the communication industry did indeed make noteworthy progress. The number of circuits per satellite increased, satellite weight per circuit decreased, antenna design improved, and frequency reuse was achieved through cross-polarization. Demand-assigned multiple access (DAMA) and time-division multiple access (TDMA) to a satellite were developed. As noteworthy as these achievements were, they were undertaken by industry only because they offered (1) a modest risk for the cost, (2) a relatively immediate market payoff, and (3) affordable development costs.

During this time period, the private sector's ability or willingness to fund high-risk, high-cost satellite communication research was limited. The only risks taken were incremental improvements for existing markets. Satellite service providers at the time required that no large investments be made unless there were reasonable assurances that very short-term payoffs would result. Providers placed a high premium on reliability and were especially averse to technology risk. Furthermore, the added peril of violating federal antitrust and trade regulations led companies to refrain from forming joint efforts that might have permitted them to share the risks. This was the environment that existed when the ACTS program was initiated.

Foreign Competition

Although the United States led the world in providing satellite communications in the mid-1970s, this lead was being contested and challenged by both the Europeans and the Japanese [10]. The advance of space industries in Europe and Japan was made possible by both direct and indirect subsidies from governments whose policies recognized the long-term importance of satellite communications. Funding for satellites in Europe and Japan continued to rise through the 1970s and into the 1980s. The real strength of the European and Japanese approach was a result of a clear focus established at the national level. The Japanese programs, including the CS series of experimental satellites, were extremely well planned, selectively focused, and designed to help Japanese industry become a major force in the world market. It should be noted that the Japanese have become the leading supplier of INTELSAT earth stations, and are a major supplier of components used in satellite communication. The objective of the European space research programs, which were entirely government-funded, was to help European industry acquire the expertise needed to compete in the world market. The European Space Agency (ESA) supported the development of the Olympus, and the Italian Space Agency supported the development of the ITALSAT flight systems that introduced critical Ka-band technology. The long-term support of these programs has resulted in viable European spacecraft suppliers for today's commercial programs (such as Globalstar, Teledesic, SES Astra 1K, and many more). During the late 1970s and early 1980s many viewed these efforts by the Japanese and the Europeans as serious threats to the U.S. lead in communication satellite development. Supporters of this view feared that U.S. communication satellite preeminence could be lost (following the unfortunate precedent in the consumer electronics industry), and they became strong supporters of NASA communication R&D. As it turns out, the Europeans began to win contracts in the late 1990s for commercial communication satellites.

U.S. Military R&D

In addition to NASA, a definite aid to the U.S. satellite communication industry is the U.S. military. There is no question that many DOD-sponsored technological developments have played an important role in the private sector's achievement of crucial technologies used in commercial satellites. Some of the military contributions have been described previously. As a more recent example, Hughes Electronics claims the work sponsored by the DOD in radiation-hardened Application Specific Integrated Circuits (ASIC), and associated digital processing has allowed them to propose their Spaceway spot beam communication system. Chapter 9 also provides additional examples. Obviously, government sponsorship of R&D can be a major positive force in the economic marketplace.

NASA's Re-entry into Communication R&D

In 1974, several organizations began to assess the consequences of NASA's decision to essentially eliminate satellite communication activities that focused on commercial applications [11]. The Electronic Industries Association (EIA) issued a position paper in January 1974, which urged NASA to reconsider its decision. In January 1975, the American Institute of Aeronautics and Astronautics (AIAA) issued a similar report. It urged NASA to re-enter the communication satellite field by sponsoring new families of application technology satellites. The report argued that from 1960 to 1973, "the federal government took the dominant role in communication satellite research and development, thereby providing the basis for low-risk operational system development by private enterprise in the 1960s and 1970s."

In the fall of 1975, NASA asked the National Research Council (NRC) to consider and report on the question: "Should federal research and development on satellite communication be resumed and, if so, what is the proper federal role in this field?" To undertake the study, the NRC formed a Committee on Satellite Communications, under the auspices of the Space Application Board. After studying this question, the consensus of the committee was that major advances in communication satellite technology *required* government investment, particularly in the areas where high technical risks were involved. This committee concluded that satellite communication R&D was an appropriate federal responsibility, and that NASA should resume the research and development activities needed to provide the new technology for future commercial communication needs. The NRC committee recommended, in a 1977 report [12], that NASA implement an experimental satellite communication technology flight program based on an assessment of need, technology projections, and service concept development. It recommended that the technical design of any NASA experimental communication satellite should support several end user service concepts, and that appropriate user groups should assist in the conceptual definition of both the needed technology and the experiments themselves.

Based on the results of the NRC report, the increasing demand for domestic voice, video, and data traffic, and the foreign competition and prospects of trade disparity, President Jimmy Carter saw fit to reinstate federal sponsorship of communication satellite technology. Official sanction for NASA to resume its responsibility was contained in the October 1978 Presidential Directive (PD-42). This directive stated, "NASA will undertake carefully selected communication technology R&D. The emphasis will be to provide better frequency and orbit utilization approaches."

The NASA Satellite Communication Program for the 1980s

In 1978, as a result of the Presidential Directive, NASA began the process of rebuilding its R&D activities in the communication satellite arena [13,14,15].

The future technology program was planned in cooperation with the National Research Council's Space Applications Board Subcommittee on Satellite Communications, whose membership consisted of leading common carriers, spacecraft manufacturers, and representatives of communication users.

Market & System Studies

In this first phase of the NASA program, market and system studies were conducted to determine future service demand and whether or not C- and Ku-band satellites could satisfy it. Two contracts were awarded to common carriers: Western Union Telegraph Company, and U.S. Telephone and Telegraph Company, which was a subsidiary of International Telephone and Telegraph (ITT) [16,17]. The emphasis of these studies was to forecast the telecommunications traffic that could be carried by satellite competitively. During this same time frame, two other system studies were conducted—one each by Hughes Aircraft and Ford Aerospace, with supporting studies by TRW, GE, and the Mitre Corporation [18,19]. Their purpose was to identify the technology needed to implement cost-effective and spectrum-conservative communication systems. The results were combined to define potential commercial system configurations that could address the market for trunking and customer premises services that was expected in the early 1990s. System requirements derived from these postulated commercial configurations formed the basis for the technology development program that followed.

The market studies predicted that rapid growth in domestic voice, data, and video traffic would lead to a five-fold increase in U.S. communication demands by the early 1990s. A combination of these market projections and communication satellite license filings with the FCC portended a saturation of North American orbital arc capacity using the C- and Ku-band frequencies.

To relieve the pressure of this expanding market, the 30/20 GHz frequency band was needed. As a result, the new NASA communication program for commercial application was named the 30/20 GHz Program and was structured to:

- Develop selective high-risk, 30/20 GHz technologies that focused on relief of orbit and frequency congestion and developing new and affordable services

- Promote effective utilization of the spectrum and growth in communications capacity

- Ensure continued U.S. preeminence in satellite communications

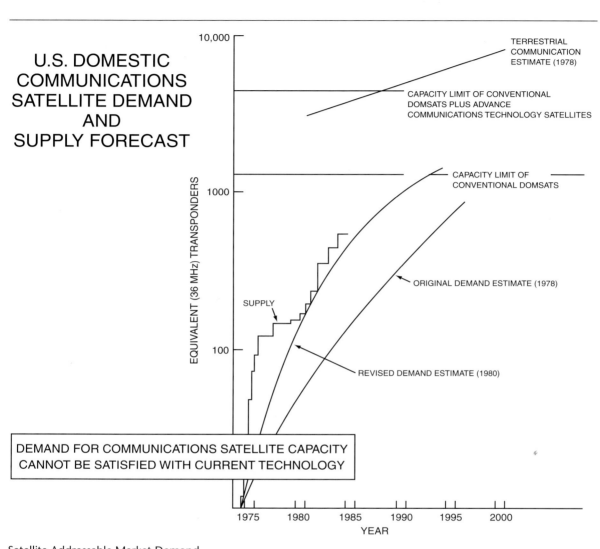

U.S. DOMESTIC COMMUNICATIONS SATELLITE DEMAND AND SUPPLY FORECAST

TERRESTRIAL COMMUNICATION ESTIMATE (1978)

CAPACITY LIMIT OF CONVENTIONAL DOMSATS PLUS ADVANCE COMMUNICATIONS TECHNOLOGY SATELLITES

CAPACITY LIMIT OF CONVENTIONAL DOMSATS

ORIGINAL DEMAND ESTIMATE (1978)

SUPPLY

REVISED DEMAND ESTIMATE (1980)

EQUIVALENT (36 MHz) TRANSPONDERS

DEMAND FOR COMMUNICATIONS SATELLITE CAPACITY CANNOT BE SATISFIED WITH CURRENT TECHNOLOGY

YEAR

Satellite Addressable Market Demand

The technologies required to meet these objectives were judged to be of such high technical risk that they were beyond the capability of any one company to finance.

In 1979, NASA designated the Lewis Research Center (LeRC) in Cleveland, Ohio, to be its lead center in planning and executing the commercial communication satellite technology R&D Program. In 1999, the Lewis Research Center's name was changed to the Glenn Research Center (GRC), in honor of John Glenn, astronaut and U.S. Senator from Ohio.

Early communication satellite systems employed simple, bent-pipe transponders with a single antenna beam to cover a large region (such as the continental United States). The new NASA program needed to develop technol-

13

ogy that would allow the frequency spectrum to be used more efficiently. One technique to accomplish this was to cover the region with many small spot beams so that the same frequency could be reused simultaneously in non-adjacent beams. Such frequency reuse increased the capacity of satellites by a factor of five to ten times that of a single beam satellite, with only a modest increase in spacecraft size, power, and weight. The technology to accomplish this high degree of frequency reuse employed antennas with high-gain spot beams and electronic systems with onboard switching and processing to interconnect the spot beams. In addition, the high-gain antenna allowed for smaller aperture user terminals at higher data rates. This was the technology developed by NASA.

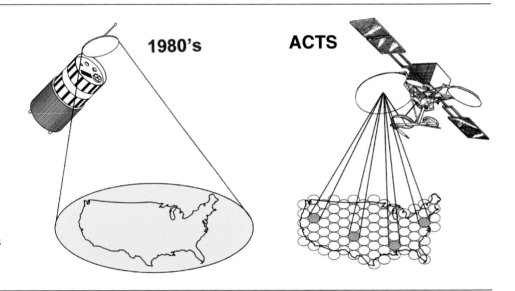

Single Beam versus Multibeam Satellites

Technology Feasibility & Flight System Definition

In 1980, the program moved forward in two phases. The first phase was to 1) continue the market studies to increase confidence in the forecast for orbit saturation and 2) to do proof-of-concept development of the identified technologies. The proof-of-concept program was a laboratory (breadboard) type of development to prove that the technologies were feasible. Approximately $50 M was expended on the first phase. If the first phase proved successful, the second phase would consist of developing an experimental flight system to demonstrate that the technologies could provide reliable communications services.

The first phase was fully supported by the entire service provider and satellite manufacturing community. The second phase of the program was the one that became controversial. The service providers had great concern about how reliably the technology would work in space, and therefore, argued for a flight program. Some satellite manufacturers, however, had reservations about proceeding with a flight program because they felt it would give the winning contractors of the NASA procurement an unfair competitive advantage. This controversy continued throughout most of the life of the ACTS program.

Program Coordination with Industry

Two industry committees were formed to guide the program. The NASA Ad Hoc Advisory Committee was created to provide general policy direction. The committee included notable representatives of both the system supplier and service supplier industry. Their contribution provided timely and sage review of the program, as well as providing NASA with insight into the industry philosophy relative to the roles and responsibilities of both government and the private sector.

NASA Ad Hoc Advisory Committee	
COMSAT	J. Harrington, Chairman 1979
American Satellite Corporation	T. Breeden
Ford Aerospace	C. Cuccia
Stanford University	C. Cutler
Digital Communications Corporation	J. Puente
RCA	J. Keigler
AT&T Long Lines	R. Latter
COMSAT	B. Edelson
MIT Lincoln Laboratories	C. Neissen
Western Union	D. Nowakoski
Member-At-Large	T. Rogers

The second industry committee was a Carrier Working Group (CWG), consisting of representatives from all the major satellite service providers. The CWG was charged with helping NASA formulate the technology and flight system requirements, develop experiments, and provide overall guidance. These requirements and experiments were deemed necessary by the CWG to demonstrate the readiness of not only the technology, but of its service applications as well.

Coordination was also established between the Department of Defense and NASA, especially in the development of various critical advanced technology components.

Carrier Working Group	
American Satellite	T. Breeden
	L. Paschall
	J. Johnson
	O. Hoernig
	F. Bowen, Jr.
AT&T Long Lines	R. Latter
	B. Andrews
Bell Telephone Labs	R. Brown
COMSAT General	J. Harrington
	J. Kilcoyne
	C. Devieux, Jr.
GTE Satellite Corporation	G. Allen
	T. Eaker
	D. Piske
	J. Napoli
Hughes Communication Services	C. Whitehead
ITT	R. Gambel
	M. Nelligan
RCA American Communications	P. Abitanto
	W. Braun
	A. Inglis
	R. Langhans
Satellite Business Systems	R. Hall
	W. Schmidt

Carrier Working Group (continued)	
Western Union	J. Spirito D. Stem D. Nowakoski
Public Service Satellite Consortium	E. Young
US Telephone and Telegraph	G. Knapp
Southern Pacific Communications	C. Waylan M. Gregory

Proof-of-Concept Development

The purpose of the proof-of-concept (POC) technology development was to demonstrate the technical feasibility of the key component building blocks [20,21]. The approach NASA used was to issue multiple contracts to various aerospace and related companies for the development of each high risk technology: *multiple spot beam antenna, base band processor, TWTA, wide-band switch matrices, low-noise receiver, GaAs FET transmitter, GaAs IMPATT transmitter,* and *ground antenna.* Duplicate awards for most of the critical technology components were employed to increase the probability of successful development, and to produce multiple sources for communication hardware. In addition, multiple awards helped to ensure that a variety of perspectives and technical approaches were brought into each development. These contracts called for the development of the technology, the construction of POC versions of the components, and their testing in the laboratory to verify performance.

The POC hardware substantially reduced the risk associated with the planned development of the flight system. Another product of these technology contracts was the prediction of feasible component, subsystem, and system performance levels. NASA used these performance predictions to provide guidance for follow-on technology development. Service providers and manufacturers could also use these predictions in planning activities for the commercial system designs of the early 1990s.

The Department of Defense (DOD) participated in the NASA POC program. Several of the critical technology POC elements that were of interest to the DOD were co-funded by DOD and NASA.

To enable the effective transfer of information that was generated in the program, all contractors were required to prepare task completion reports.

17

These reports were presented at periodic industry briefings (only for interested U.S. parties) hosted by NASA.

Flight System Definition Studies

The need for a flight test program reflected the fact that much of the required technology had never been demonstrated in space. The flight test was to ensure that the technology base was mature and validated, providing the level of confidence recommended by industry as being necessary for commercial exploitation. The initial planning called for two experimental satellites to be built and flown; one to demonstrate telephone trunking for high volume users in metropolitan areas, the second to demonstrate customer premises services using small and inexpensive earth terminals located at customer locations. In 1980, the two-flight concept was reduced to a single experimental spacecraft, primarily emphasizing customer premises services. This proved to be a wise decision since the introduction of fiber optics a few years later significantly reduced the cost for terrestrial trunking services, making satellites non-competitive.

In February 1982, Dr. Burt Edelson became NASA's associate administrator of the Office of Space Science and Application, and played a very important role in keeping the program alive. When the program was seriously threatened in 1982, Dr. Edelson restructured the 30/20 GHz program by broadening its applicability to the entire frequency spectrum for satellite communications. As a result, the experimental satellite system was renamed the Advanced Communications Technology Satellite (ACTS), and it focused primarily on the technology of multi-beam antennas and associated onboard switching and processing. Spacecraft capacity was reduced to a minimum for technical verification and experimentation only. Dr. Edelson provided key leadership for the ACTS program during his tenure at NASA, and has been a vocal proponent of the program and its benefits ever since.

Two other NASA managers who provided important leadership to the NASA Communications Program were Joe Sivo and Bob Lovell. Joe Sivo was the chief of the Communications and Applications division at NASA's Lewis Research Center in Cleveland, Ohio. Joe was the "Father of ACTS" and led the LeRC team in the late 1970s and the early 1980s as NASA restructured its communication program. Bob Lovell became chief of the Communications division at NASA Headquarters and worked with both Dr. Edelson and Joe Sivo to structure the ACTS program and guide it through technical and political hurdles in the early 1980s. Without Sivo, Lovell, and Edelson, the ACTS flight program would have never gotten off the ground.

Dr. Burt Edelson
with Joe Sivo (left)
and Bob Lovell
(right).

Concurrent with the POC technology development, NASA was working with industry to define flight system concepts that would demonstrate ACTS technology readiness and its service capabilities. During the period of 1981-1983, the major spacecraft manufacturers—Ford Aerospace (now Space Systems Loral), Hughes Aircraft, TRW, GE, and RCA (both now part of Lockheed Martin) were funded by NASA to conduct system studies for defining a R&D spacecraft (ACTS) that could be flown by NASA. NASA then used the results from these system studies to develop the Request for Proposal (RFP) for the ACTS spacecraft and ground system. This RFP was issued by NASA in early 1983, with a proposal due date of June 1, 1983. Since the RFP required the development of very high-risk technology that had never been flown before, a cost-plus-fixed-fee type of contract was specified.

The five separate flight system studies were conducted to get a wide range of views on what the ACTS spacecraft configuration should be and to promote competition for the procurement of the spacecraft. As it turns out, this process did not accomplish the latter objective and was complicated by the fact that there was not a clear consensus for the need for a flight program to prove the feasibility of the new technology.

The Reagan administration espoused a minimal government involvement ideology. At the time, the Republican administration took the position that it was not the proper role of government to conduct a flight program for the purpose of proving technology for commercial purposes, especially for a profitable industry. There were many arguments presented by the Republican administration as to why the government should not sponsor the flight verification. These included arguments that the government was not capable of predicting technology for commercial application, and that the spot beam, frequency-reuse technology was not necessary because there was plenty of C- and Ku-band spectrum for future use. However, as we know today, the use of spotbeams allows a great increase in the amount of frequency reuse so that a single satellite can have a very large capacity. Without this spot beam increase in capacity, many of the mobile and broadband satellite systems under development in the late 1990s, such as Iridium, Globalstar, Spaceway, Astrolink, and iSKY (formerly called KaStar), would not have been economical. All the developers of these systems make a strong case that their spot beam systems meet the current FCC requirement to more efficiently use the spectrum. The FCC has added this requirement since they realized that the frequency spectrum is a scarce resource.

Congress in the 1980s was increasingly concerned about U.S. economic competitiveness in high technology industries. They were sensitive to areas such as satellite communication being challenged by foreign entities, where the federal government could improve U.S. competitiveness. The Democratic Congress listened to the arguments of the U.S. commercial satellite industry in support of a flight verification program and decided it should be conducted. This debate between the Republican administration and the Democratic Congress (including each side's constituencies) over the need for the ACTS flight test continued through launch of the ACTS in September of 1993. Later chapters in this book cover this debate in much more detail. It is sufficient here to say that the difference in philosophy between the White House and Congress was great enough that Reagan's budget left the program without funds for five years in a row, and that Congress restored the funds in each budget during those five years. If nothing else, the ACTS program may have set a record in this regard.

Bidding on the ACTS Contract

The response to the ACTS RFP was disappointing because only one proposal was received. The team submitting the proposal consisted of RCA as prime integrating contractor and supplier of the spacecraft bus, with first-tier subcontractors TRW (for the communication payload) and COMSAT (for the Master Control Station). Second-tier subcontractors included Motorola for the base band processor and Electromagnetic Sciences (now called EMS Technologies) for the spacecraft's antenna beam-forming network. Since TRW, Motorola, and Electromagnetic Sciences had developed major pieces of the ACTS technology in the proof-of-concept development program, this team was very competent. Because the team represented a large cross-section of the U.S.

20

industry involved in satellite communications, NASA believed that objectives of the program could still be achieved by the single bid.

ACTS was to be placed into a low earth orbit by NASA's space shuttle, and the RFP required that the payload be constrained to as small a space in the shuttle's cargo bay as possible. One option was to use a Payload Assist Module PAM-D perigee stage to place ACTS into a geostationary transfer orbit (GTO) after deployment from the shuttle. Some pre-proposal studies showed that this approach would result in the ACTS only using one quarter of the shuttle's cargo bay volume. The next alternative was to use a larger capacity perigee stage—a PAM-A—which would take up more volume in the cargo bay. NASA wanted to restrict the payload's volume in the cargo bay to limit the total cost of the mission, including launch. At this time, shuttle launch costs were based on the volume occupied in the cargo bay. This logic was somewhat questionable since the shuttle was not always flown with a full load. In fact, the NASA cost model used a shuttle load factor of 3/4 capacity to determine the pricing for payloads. Potential bidders questioned this requirement and sought a change. Prior to the receipt of proposals on June 1, 1983, Robert Berry, director of Space System Operations at Ford Aerospace, wrote to NASA on May 3, 1983, [22] and stated the following:

> We believe that the technical approach which would create the least risk to NASA would be a PAM-A configuration satellite, unconstrained by the volume limitations of the PAM-D family of perigee stages. We have made the case that it is in NASA's best interest not to discourage the offer of a PAM-A configuration. It seems obvious to us that NASA's objectives in achieving an ACTS program offering innovative and unique associations with other government or commercial users can only be satisfied with a PAM-A equivalent spacecraft. We further believe that as the definition of NASA's high technology payload evolves, weight, power, footprint area, thermal considerations, and performance margins will move toward the limitation of the PAM-D configuration. On the other hand, ample margin would still exist with a PAM-A configuration.
>
> Ford Aerospace had planned to offer a PAM-A class spacecraft for the ACTS mission. We also had received notification from another payload user of their firm commitment for another payload, which would have been incorporated, on our satellite configuration along with the ACTS payload. In addition, we had been informed by a satellite operating company of its interest in leasing the ACTS payload, thus providing potential cost reimbursement to NASA. However, RFP 3-511907 quantitatively defines the assignment of launch costs to overall program costs, but offers no quantitative offset for the substantially greater capability of a PAM-A configuration. This quantitative imbalance confers an apparent competitive cost advantage to a PAM-

21

D class satellite configuration even though that configuration will not support the full achievement of NASA's overall program objectives.

Should NASA subsequently decide that a PAM-A equivalent configuration is desirable for ACTS, Ford Aerospace would be pleased to offer a competitive solution.

In a December 1983 *Aviation Week & Space Technology* article [23], Berry went further and said, "There is no way the ACTS payload is compatible with the McDonnell Douglas Delta (PAM-D) class upper stage."

It was expected that the ACTS contractor would use a standard commercial bus to limit the non-recurring costs for the spacecraft. Ford Aerospace wanted to bid its standard PAM-A bus, which evidently would have taken up considerable volume in the shuttle. They must have perceived that this would have made them non-competitive, so therefore did not bid. As it turned out, Berry's statement that the ACTS weight requirement was beyond the PAM-D capability was true. When RCA bid the ACTS job, they proposed a PAM-A configuration with the antenna reflectors folded across the top of the spacecraft to minimize the volume taken up in the shuttle. Not long after the contract was awarded, the folded reflector design was replaced with a truss structure arrangement that significantly increased the needed volume in the cargo bay. When ACTS was launched, it took up approximately one half of the cargo bay's volume. NASA's concern with limiting shuttle costs and its procurement regulations forbidding informal discussions with potential bidders after release of the RFP, resulted in improper treatment of a contractor's input. In hindsight, it is obvious that it would have been better for the program had the shuttle launch costs not been included in the proposal evaluation. The result would have been the receipt of two proposals instead of one.

The other major potential bidder for ACTS was Hughes Aircraft Company. Although the reasons were not publicly stated, Hughes chose not to submit a bid for ACTS.

Hughes Ka-band Filing

Hughes questioned the ACTS program in principle as unnecessary subsidization of commercial operations and duplication of military technology development. To emphasize their point, they filed an application with the FCC in early December of 1983 for the development, launch, and operation of a two-satellite Ka-band domestic system. Their satellites were to be equipped with high-power spot beams focused on 16 major U.S. metropolitan areas. As such, it would allow the use of two meter customer-premises earth stations for business data services such as teleconferencing, high-speed document distribution, and remote printing. The first of their two proposed satellites was to be launched in December of 1988.

Hughes noted in this filing that they expected the orbital allocations at the C- and Ku-bands to be exhausted following the next round of FCC assignments. In essence, Hughes agreed with NASA's C- and Ku-band saturation projections, which had been derived by Western Union. In fact, Hughes quoted the Western Union market study in their filing. This filing, however, was contrary to other statements made by Hughes during the same time period that warned of a coming glut in transponder capacity. This contradiction and other factors led to speculation by observers [5] that Hughes opposed the ACTS flight program on purely competitive grounds. Nothing ever came of the Hughes filing, but it did set the stage for consistent Reagan administration opposition to ACTS.

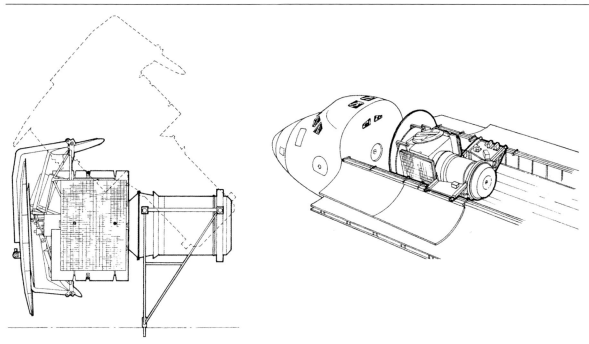

ACTS spacecraft attached to PAM-A stage for mounting in the space shuttle. Dashed line indicates elevated position for deployment from the shuttle. (Figure taken from the RCA Astro Space proposal received by NASA on June 30, 1983.)

The administration turned against continuing ACTS as a flight program after Hughes filed with the FCC. Since ACTS and Hughes' system used the same 30/20 GHz frequencies, senior Hughes executives argued against continuing the program before the Office of Management and Budget (OMB) and NASA, claiming that a government-funded program would be redundant.

When the Reagan administration sent its budget proposal for NASA to Congress at the end of January 1984, it reduced the funding to so small a level that a flight program could not proceed.

ACTS Contract Signed

In the ensuing months, a battle over the ACTS flight program was waged in the U.S. Congress. Congress became convinced that the ACTS program objectives were valid and important to carry out. In the latter part of May 1984, despite administration opposition, they approved a $40 M increase for the ACTS program to reinstate the flight verification phase. After President Reagan signed the FY 1985 authorization and appropriation bills, it cleared the way for a contract signing with RCA on August 10, 1984, for the development of the ACTS flight system. This funding battle over ACTS between the administration and Congress continued for the next four years, with the administration trying to terminate ACTS each year.

As initially proposed by RCA, ACTS system development was to take place over a five-year period with an engineering model development being completed in three years. Because of the complex coordination between the user terminals, the master control station, and the onboard switch system in setting up on-demand circuits, the development included a comprehensive, three-month test of the ACTS ground system with the spacecraft. The proposed five-year development time contrasts with the normal commercial satellite development of three years, and reflects the fact that the ACTS technology was well beyond the current state-of-the-art. With the ACTS contract awarded in August of 1984, the scheduled launch date was September of 1989. As described in a later chapter, funding cutbacks, development problems, and other difficulties caused the launch to be delayed until September of 1993.

Changing Times

What is the proper role of government in technology development? The NASA ACTS program served a very important role in advancing satellite communication because the commercial satellite communication industry in the 1980s could not afford to take on the risk associated with the necessary technology.

The business climate in the 1980s was entirely different than in the late 90s. Today, non-traditional satellite companies such as Motorola, iSKY, and Pasifik Satelit Nusantara (PSN)—to name just three—have found investor partners to put up billions of dollars for implementation of revolutionary satellite communication systems employing advanced technologies. Iridium is a 66

LEO satellite system that provides mobile communication anywhere in the world. iSKY is a GEO satellite system that provides broadband communications for the consumer Internet in the United States. In the case of PSN, the system is a GEO, handheld mobile communications system called ACeS. All three systems use multiple spot beam antennas, and both Iridium and AceS have on-board digital processing. Iridium also has inter-satellite links to provide global connectivity. Because of the success of many new satellite systems—such as NASA's ACTS, DOD's MilStar, and Hughes' DirecTV—many satellite service providers now view new technology not as a major risk factor but as a means to introduce new services. Another major difference today is that the perceived market potential is *much* greater than it was in the 1980s.

In the 1980s, communications satellites were still in their infancy and large sums of capital were not available for risky ventures. The ACTS flight program was a proper role for the government in the 1980s. Due to differences in the business climate and the maturity of many technologies, a similar flight program is not considered necessary today. This is discussed in more detail in Chapter 9, "The Role of Government in Technology Development."

CHAPTER 2

SATELLITE TECHNOLOGY

Communication satellites provide a unique perspective with which to view the earth's surface. At the geostationary altitude of 22,240 miles above the equator, they appear motionless in the sky. Serving as giant relay towers, they interconnect users in vast areas of the world who are within their continental field of view.

Due to interference considerations, communication satellites must maintain a certain separation. Therefore, only a limited number of satellites can be placed in geostationary orbit to provide communications for a region such as the United States. In addition, only certain radio frequency bands, assigned by international agreement, are available for commercial communication satellite use. The extraordinary success of satellite communication in the late '70s and early '80s threatened to exhaust both the available frequencies and the geostationary satellite positions for many regions in the world. New technology was needed to provide for this projected increase in demand. In addition to finding ways to use the existing frequency bands more efficiently, operations in the next higher frequency band (the Ka-band) were deemed necessary.

NASA's ACTS program provided new technology for increased efficiency using all radio frequencies including Ka-band. Increasing the spectrum efficiency was achieved by developing high-gain, multiple spot beam antennas and onboard switching and processing that allowed for a great increase in the number of times the same frequency could be reused by a single satellite [24]. In addition, the high-gain spot beams provided the very desirable benefit of allowing for smaller aperture user terminals at higher data rates.

NASA and the U.S. commercial satellite industry jointly defined the ACTS program. ACTS was not intended to be an operational system. It was designed to be a test bed for verifying those advanced technologies that were beyond the ability of any one satellite company to finance. In the early 1980s, the U.S. satellite carriers had great concerns about the reliability of ACTS' advanced, high-risk technology. Companies felt that a flight test was necessary to prove the technology was feasibile before they would incorporate it into their commercial systems. The ACTS program was designed to allow U.S. industry the opportunity to meet the communication needs of the twenty-first century while remaining competitive in the international satellite communication marketplace. The motivation for the program and its merits are discussed in Chapters 1, 7, 8, and 9. This chapter describes the technological advances made by the ACTS program.

ACTS System Overview

ACTS is an in-orbit, advanced communication satellite test bed, bringing together industry, government, and academia in a wide range of technology, propagation, and user application investigations. NASA's Lewis Research Center (LeRC) awarded the ACTS contract in August of 1984 to an industry team consisting of:

- Lockheed Martin, East Windsor, New Jersey – for system integration and the spacecraft bus

- TRW, Redondo Beach, California – for the spacecraft communication payload

- COMSAT Laboratories, Clarksburg, Maryland – for the network control and master ground facility

- Motorola, Chandler, Arizona – for the baseband processor

- EMS Technologies (formerly called Electromagnetic Sciences), Norcross, Georgia – for the spot beam forming networks

The contract was actually awarded to RCA Astroelectronics of East Windsor, New Jersey (which was subsequently acquired by General Electric (GE), then by Martin Marietta, and is currently part of Lockheed Martin). In 1988, as a result of a congressionally mandated program funding cap, Lockheed Martin (General Electric Astro Space at that time) assumed responsibility for completing the development of the communication payload. Subsequently, Lockheed Martin subcontracted with Composite Optics, Inc., in San Diego, California, for the manufacture of the antenna reflectors and part of the bus structure.

ACTS was launched into orbit by the space shuttle Discovery (STS-51) on September 12, 1993, and achieved geostationary orbit at 100 degrees west longitude on September 28, 1993. As of the printing of this book, ACTS is still operating, but the in-orbit stationkeeping fuel has been depleted. Operations continue with an inclined orbit, using an autonomous, onboard program that provides a bias in the roll axis to offset the inclination and maintain the spot beams properly located on the ground.

The ACTS system is made up of a spacecraft and ground segment [25-28]. The spacecraft consists of a multi-beam communication payload and the spacecraft bus. The key technological components of the communication payload are the multi-beam antenna (MBA) assembly, the base band processor (BBP), the microwave switch matrix (MSM), and Ka-band components. The spacecraft bus houses the communication payload and provides attitude control, electric power, thermal control, command reception, telemetry transmissions, and propulsion for stationkeeping.

A NASA ground station (NGS) and master control station (MCS), collocated at the Lewis Research Center (LeRC) in Cleveland, Ohio, transmit commands to the satellite, receive all spacecraft telemetry, perform ranging operations, and provide network control for all user communication. The NGS/MCS process and set up all traffic requests and assign traffic channels on a demand basis. A satellite operations center was located at Lockheed Martin Astro Space in East Windsor, New Jersey, and connected to the NGS/MCS via landlines.

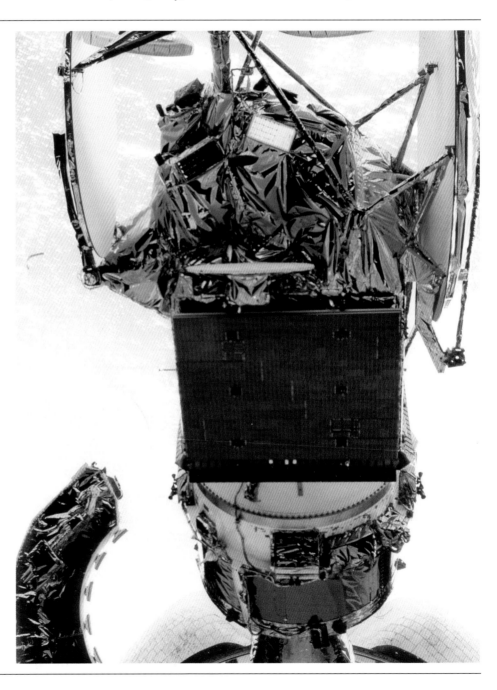

ACTS deployment from space shuttle Discovery (STS-51) on September 12, 1993. Bottom to top: Orbital Sciences Transfer Orbit Stage (TOS), folded solar panels against spacecraft bus, and ACTS antenna assembly with main reflectors folded for launch. Lockheed Martin headed the contractor team that designed and built the ACTS. The TOS burn was well within 1 sigma performance.

In June of 1998, the Satellite Operations Center was transferred to the Lockheed Martin Communications and Power Center facility in Newtown, Pennsylvania. The Satellite Operations Center has the prime responsibility for generating spacecraft bus commands and for analyzing, processing, and displaying bus system telemetry data. Orbital maneuver planning and execution are also handled by the Satellite Operations Center. The Lockheed Martin C-band command, ranging, and telemetry station at Carpentersville, New Jersey, provided transfer orbit support during launch and originally served as an operations backup to the Satellite Operations Center. In 1998, however, the backup function was transferred to the GE American Communications station in Woodbine, Maryland.

The ACTS communication payload provides service at digital data rates from kilobits per second (Kbps) up to hundreds of megabits per second (Mbps) via its various communication modes of operation. The major types of services include:

- On-demand, integrated voice, video and data services using T1 (1.544 Mbps) links to 4-foot customer-premises terminals

- Very high-data-rate (622 Mbps) networks

- Broadband (T1) video and data for aircraft and ships

- Aeronautical voice and low rate data

- Low-rate terrestrial mobile voice, video, and data

- Interactive, multimedia services (1 Mbps outbound and 20 Mbps inbound) using 18-inch terminals

Terminals operated by various private, governmental, and university organizations validated these services. In addition, more than ten receive-only propagation terminals were used for propagation studies and modeling.

ACTS is a three-axis-stabilized spacecraft that weighed 3,250 pounds at the beginning of its in-orbit life. It measures 47.1 feet from tip to tip along the solar arrays, and 29.9 feet across the main receiving and transmitting antenna reflectors. The ACTS multi-beam antenna is comprised of separate Ka-band receive and transmit antennas, each with horizontal and vertical polarization subreflectors. Antenna feed horns produce narrow spot beams with a nominal 120-mile coverage diameter on the surface of the Earth. Fast (less than 1 microsecond [μsec]), beam-forming switch networks consisting of ferrite switches, power dividers and combiners, and conical multi-flare feed horns provide sequential hopping from one spot beam location to another. These hopping spot beams interconnect multiple users on a dynamic, traffic-

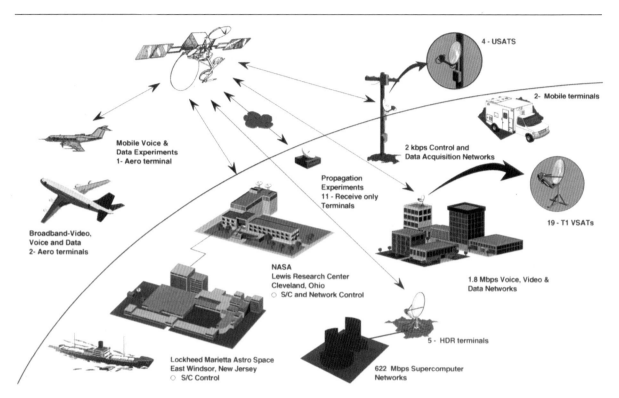

Overview of ACTS system showing spacecraft, control stations, and user terminals. Lockheed Martin operates the control facility in New Jersey while Comsat Laboratories performs operations in Ohio.

demand basis. A separate 3.3-foot, mechanically-steered antenna, receiving uplink and radiating downlink signals, is used to extend the ACTS communication coverage to any location within the hemispherical field of view from ACTS' 100 degree west longitude position. Beacon signals at 20.2 GHz and 27.5 GHz are radiated from two small, separate antennas.

Communication Modes of Operation

The multi-beam antenna provides dynamic coverage with fixed and hopping spot beams. Each hopping spot beam can be programmed to sequentially cover a set of spots and dwell long enough to communicate with users in each spot. By assigning each user an access time, several users can transmit and receive at the same frequency on a time-shared basis. This time division, multiple access (TDMA) technique requires a switching system onboard the spacecraft to interconnect the beams and route messages. The ACTS communication payload provides two types of onboard switching to interconnect the

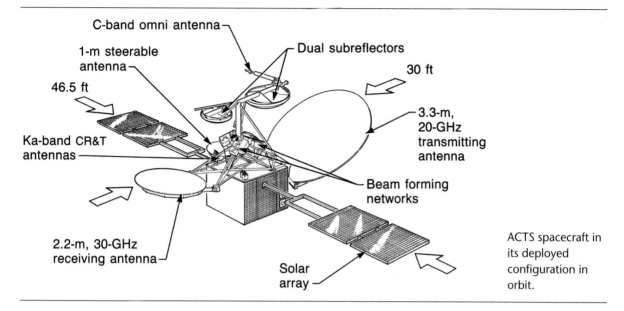

C-band omni antenna

1-m steerable antenna

46.5 ft

Dual subreflectors

30 ft

3.3-m, 20-GHz transmitting antenna

Ka-band CR&T antennas

Beam forming networks

2.2-m, 30-GHz receiving antenna

Solar array

ACTS spacecraft in its deployed configuration in orbit.

multiple spot beams and route signals to their appropriate destinations: 1) base band processing (BBP) and 2) microwave switch matrix (MSM).

The BBP is a high-speed digital processor on the satellite that provides on-demand, circuit switching for the efficient routing of traffic among small user terminals. In essence, the BBP is the first switchboard in the sky to perform the same functions done by terrestrial telecommunication switch centers. Because its network is completely interoperable with the terrestrial system, ACTS can be considered a single node in a combined satellite/terrestrial network.

ACTS conducts both time and space switching on board the satellite. The BBP switches traffic between the various uplink and downlink beams, automatically accommodating on-demand circuit requests. In the BBP mode of operation, four simultaneous and independent hopping beams (two uplink and two downlink) provide flexible, demand access communication between small (4-foot diameter antenna) user terminals with a maximum throughput of 1.79 Mbps or 28 64-Kbps circuits. Each uplink spot beam receives multiple channels.

A user terminal is assigned an uplink channel and transmits its information using Time Division Multiple Access (TDMA). At the spacecraft, the receive signals are demodulated, decoded as required, temporally stored in memory, routed on a 64-kilobit individual circuit basis, modulated, encoded if required, and transmitted in the proper downlink spot beam using a single TDMA channel. During the 1 millisecond TDMA frame time, the beams hop to many locations, dwelling long enough to pick up or deliver the required traffic.

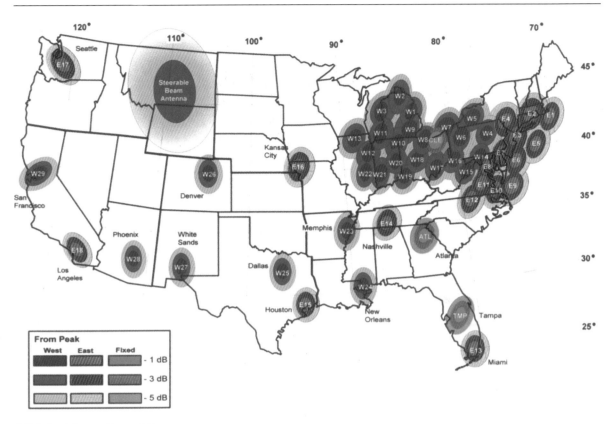

ACTS Spot Beam Ground Coverage

The MSM is an intermediate frequency (IF) switch capable of routing high volume point-to-point traffic and point-to-multipoint traffic over 900 MHz bandwidth channels. Using satellite-switched TDMA, the microwave switch matrix dynamically interconnects three uplink and three downlink beams. The user terminals transmit TDMA bursts according to their destination. At the satellite, the 30 GHz bursts are down-converted to an intermediate frequency, routed to the proper downlink beam port, up-converted to 20 GHz, and transmitted on the downlink. The switch paths are changed during guard intervals between bursts. Fixing the beam interconnections in a static mode allows additional flexibility for a variety of continuous digital or analog communication. The MSM mode accommodates user terminals operating from low kilobits per second up to 622 megabits per second.

The ACTS system can be configured in the BBP mode, the MSM mode, or a mixed mode. In the mixed mode, both the base band processor and the microwave switch matrix are operated simultaneously with some restrictions. The system can be quickly reconfigured from one mode of operation to

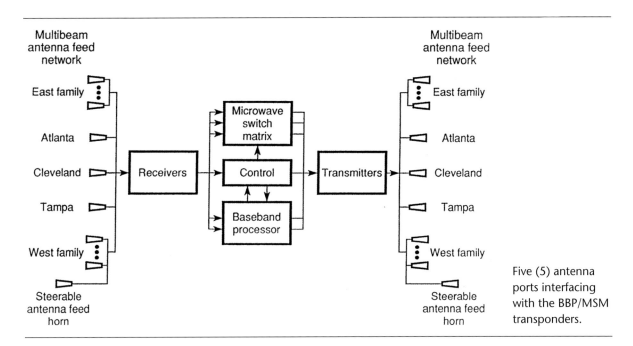

Five (5) antenna ports interfacing with the BBP/MSM transponders.

another in a matter of minutes, further adding to the system's flexibility. This flexibility, along with the large total information throughput capacity, allows a large variety of users to be accommodated concurrently.

ACTS Payload Components

The key technologies developed and flight-tested by ACTS are the multi-beam antenna, the base band processor, and the microwave switch—along with RF components operating at the Ka-band frequency. The next sections in this chapter provide a comprehensive technical description of each of the major technology components of the ACTS payload.

Multi-beam Antenna (MBA)

The ACTS multi-beam antenna (MBA) provides very high-gain spot beams, enabling the use of smaller aperture user terminals and increasing the amount of frequency reuse. In the early 1980s, traditional C- and Ku-band satellites reused their frequency spectrum by utilizing cross-polarization. The ACTS MBA provides for frequency reuse, not only by cross-polarization but also by spot beam, spatial isolation. For ACTS, spots separated by one beam width can use the same frequency and polarization. Using this ACTS approach, commercial multiple spot beam systems being constructed in the late 1990s typically achieve a frequency reuse factor of 8. Increasing the frequency reuse allows

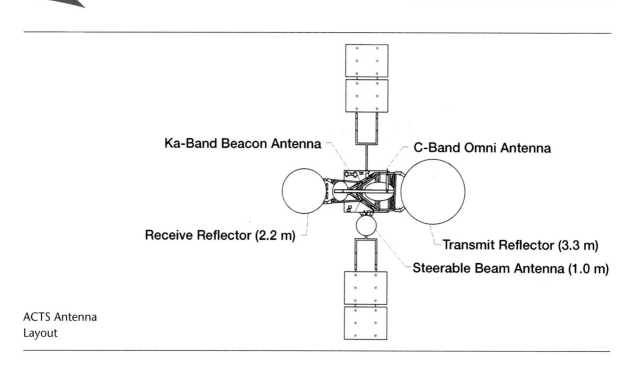

Ka-Band Beacon Antenna

C-Band Omni Antenna

Receive Reflector (2.2 m)

Transmit Reflector (3.3 m)

Steerable Beam Antenna (1.0 m)

ACTS Antenna
Layout

the total communication capacity of a single satellite to be much greater. With a larger capacity per satellite, the economies of scale are such that the satellite cost per unit bandwidth are reduced by a factor of 2 or more.

The ACTS antenna system is comprised of a multi-beam, high-gain, hopping spot beam antenna, a separate mechanically-steerable beam antenna (SBA), and two small beacon reflectors [29-31]. A C-band omnidirectional antenna, which provides tracking, telemetry, and command during the launch phase, also serves as backup during in-orbit operations.

The multi-beam antenna consists of separate receive (30 GHz) and transmit (20 GHz) antennas. Each antenna consists of a main reflector, a nested assembly of one front and one back subreflector, and a pair of feed assemblies—one each for horizontal and vertical polarization. This Cassegrain-type antenna was chosen to provide a compact design with a large focal length to minimize scan losses at the edge of the continental United States (CONUS). A large, equivalent focal length to aperture diameter was achieved using a hyperbolic subreflector with a high magnification factor. The MBA was designed so that the receive beam for a given coverage location was orthogonally polarized to the transmit beam in the same location. In-orbit, the spacecraft is oriented so the receive reflector is on the east side and the transmit reflector is on the west side.

The 20 GHz transmit antenna reflector is 10.8 feet in diameter and the 30 GHz receiver antenna reflector is 7.2 feet. This inverse scaling of reflector size with frequency produces spot beams having the same angular size in both the

20 and 30 GHz frequency bands, thus providing the same ground coverage footprint. Both antennas produce spot beams with approximately 0.3 degree beam width and gains ranging from 49 to 55 dBi. Spot beam ground coverage is on the order of 120 miles in diameter. For contract reasons, the ACTS MBA was configured to be a completely separate physical system that is bolted on the top of the bus at only three points. Although this simplified the contractor interface, it resulted in an extra heavy antenna assembly. Making the antenna an integral part of the bus would have resulted in a much lighter system. Commercial satellites take this lighter-weight approach.

Since ACTS is a test bed, it is not intended to provide full U.S. coverage. One of its prime goals is to demonstrate frequency reuse by means of hopping spot beams in two contiguous sectors. A contiguous coverage area of approximately 20 percent of the U.S. was selected. In order to further enhance the versatility of the ACTS system and provide coverage to users in other parts of the country, isolated beams outside these two contiguous areas were identified and incorporated into the hopping beam coverage network.

Each transmit and receive antenna supports five separate antenna ports. These five ports are for the three stationary fixed beams focused on Cleveland, Atlanta, and Tampa, plus the east and west hopping spot beam families. The east beams are orthogonally polarized to the west beams in both uplink and downlink while using the same Ka-band frequency. The east family of beams is comprised of a contiguous sector called the east scan sector plus six additional isolated spot locations outside the sector. The west family comprises a west scan sector plus seven additional isolated spot locations. The mechanical SBA is also interconnected into the west family and functions as part of its hopping beam network. With this antenna system, ACTS can provide service to 51 separate spot beam locations (see figure on page 34).

To prove the feasibility of the antenna design, TRW fabricated and tested a breadboard multi-beam antenna assembly. The main reflector was 8.9 feet in diameter. Testing on an indoor range verified that the antenna radiation patterns and scan characteristics were as predicted. It was found that the side lobe levels were not significantly affected by the surface distortion of the subreflectors. In addition, the front subreflector only negligibly affected the side lobes from the beams formed by the rear subreflector. However, the use of dielectric support ribs behind the gridded front subreflector introduced amplitude and phase distortions. These distortions resulted in a badly distorted far field radiation pattern. TRW concluded that the front subreflector must have no ribs. Later, this requirement proved to be a major mechanical design problem and resulted in in-orbit beam wandering. The successful demonstration of the breadboard MBA design provided the necessary confidence to initiate the development of the ACTS flight model MBA.

Plaster mold for main reflectors which were manufactured by Composite Optics.

Main Reflectors The reflectors were light, stiff, and strong, with a root mean square surface (rms) accuracy of 0.003 inches. They were thermally stable over large daily temperature gradients. The main reflectors, which were made of graphite epoxy skins bonded to a honeycomb core, were formed on a precision mold with a surface accuracy of 0.001 inches. Six-ply graphite epoxy front and back face sheets were bonded to a Kevlar honeycomb core. This reflector shell was supported by a backing rib structure with an egg crate design. The 10.8-foot, 20 GHz transmit and the 7.2-foot, 30 GHz receive reflectors weighed about 80 and 30 pounds, respectively.

Subreflectors Each subreflector assembly consisted of two offset hyperbolic surfaces arranged in a piggyback configuration. The front surface was a copper grid passing one polarization and reflecting the orthogonal polarization. The focal axis of each surface was tilted either plus or minus 10 degrees from the symmetry plane of the main reflectors, so that the two orthogonal polarized feed assemblies could be placed in their respective focal regions without mechanical interference (see figure on page 41). The solid back surface was similar in design and construction to the main reflectors, which consisted of a graphite composite/Nomex honeycomb shell with supporting back ribs. The construction of the front surface was very different, being formed from Astroquartz sheets bonded to a Nomex core. Each shell had a thickness of approximately three-quarters of a wavelength at the frequency of operation. The polarizer grid consisted of an etched copper clad on a 0.0005 inch Kapton

Supporting rib structure for 7.2-foot, 30GHz main reflector.

material that was bonded to the front face sheet. A support ring bonded to the shell around the perimeter reinforced the front subreflector. As it turned out, this ring was not adequate to prevent thermal distortions that produced significant diurnal spot beam drift. The two-way measured insertion loss through the polarizer grid was less than 1 dB. Reflection loss was measured at less than 0.25 dB.

Both the main and subreflector graphite fiber-reinforced composite surfaces had high electrical conductivity and did not exhibit significant loss at the Ka-band frequencies. To limit temperature extremes, the surfaces of all reflectors and subreflectors were coated with a low-insertion-loss thermal paint. In addition, the rear surfaces of all reflectors were blanketed for thermal control.

Lockheed Martin designed (and Composite Optics fabricated) all of the MBA flight reflectors except the steerable beam antenna, which was manufactured by Dupont Chemical. Although Composite Optics had already built 50 space-based reflectors and deployed 30 of them by the time ACTS was developed in 1988, the ACTS reflectors were a particular challenge because of their large size and stringent surface accuracy requirements. Composite Optics found that the tooling techniques used to develop antenna molds were criti-

20 GHz transmit main reflector weighting 80 pounds with a 10.8-foot diameter and a rms surface accuracy of 0.003".

Manufactured by Composite Optics.

cal in achieving the required high surface accuracy. In addition, the amount and area coverage of adhesive used to bond the outer skins to the inner honeycomb was critical in achieving reflectors that would not delaminate under the large thermal gradient expected in-orbit.

Feed Assemblies & Beam-Forming Networks The receive and transmit antennas each operated with two feed assemblies (four total). The two feed assem-

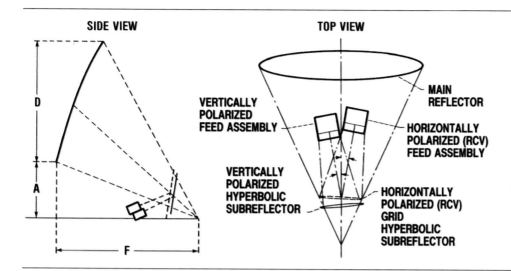

SIDE VIEW TOP VIEW

D
A
F

VERTICALLY POLARIZED FEED ASSEMBLY

VERTICALLY POLARIZED HYPERBOLIC SUBREFLECTOR

MAIN REFLECTOR

HORIZONTALLY POLARIZED (RCV) FEED ASSEMBLY

HORIZONTALLY POLARIZED (RCV) GRID HYPERBOLIC SUBREFLECTOR

Layout of the main reflector, the two subreflector surfaces, and the two feed assemblies. Layout applies for either the transmit or receive antenna.

blies for the transmit and receive antennas had orthogonal polarization and contained a total of 51 feed horns. Three of these horns were driven directly and provided fixed spot beams at Cleveland, Atlanta, and Tampa. The remaining feed horns were driven by two beam-forming networks (BFN) and provided the hopping spot beams. These feed arrays were located in the focal plane of the reflector/subreflector system for the designated polarization. The ACTS BFNs, built by EMS Technologies [32], consisted of three-way power dividers or combiners, sections of waveguide, and ferrite switches—implemented as a switching tree (see figure on page 43). Using the ferrite circulator switches, single feed horns were activated for the isolated spot beams while feed horn triplets were activated for the sector spot beams. The beams were switched in less than 800 nanoseconds (ns). These switches were interconnected by gold-plated, electroformed, copper waveguide. Manufacturing tolerances of the BFNs were tightly controlled to assure that the three activated ports for each feed triplet were phase-matched to within a few degrees over the receive and transmit frequency bands. These ferrite switches had a low insertion loss of approximately 0.1 dB, low mismatch loss, and better than 18 dB isolation across a 1 GHz bandwidth. The beam-forming networks with electronics but without feed horns ranged in weight from 25 to 41 pounds. Using 1990s technology, EMS Technologies has stated that these weights could be reduced to 1/3 of these values. The RF losses through the beam-forming networks were less than 0.6 dB for isolated spots and less than 1.2 dB for the scan sector spots. In the BBP mode of operation, a maximum of 48 hopping-beam locations can be interconnected, and communication traffic can be exchanged between locations over a frame period of 1 millisecond. This

41

process is repeated every millisecond, enabling continuous communication between locations.

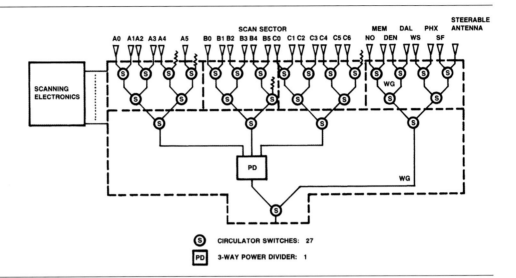

RF switch tree for the west beam-forming network.

The feed assemblies incorporated three different feed horn types [33]. Multi-flare horns with a 17 dB feed taper for low side-lobe performance were used for the Cleveland, Atlanta, and Tampa fixed spots. The isolated spots of the hopping beams used multi-flare horns with a 10 dB feed taper for maximum peak gain. Multi-flare horns were selected because they generate circularly symmetric radiation patterns and had a lower manufacturing cost compared to the dual mode or corrugated feed horns. The horns used in the triplet configurations were simple, single flare horns with a 3 dB feed taper. The multi-flare horns were manufactured of lightweight aluminum while the smaller, single flare horns were fabricated of electroformed copper.

Since ACTS was a test bed, the antenna focal planes were not fully populated with feed horns to provide complete CONUS coverage. The design, however, could be readily scaled for such coverage [31].

Steerable Beam Antenna (SBA) The ACTS system also incorporated a separate, mechanically steerable beam antenna whose single reflector was used for both uplink and downlink. The SBA complemented the MBA by providing additional coverage anywhere in the Western Hemisphere as viewed from ACTS' 100 degree west longitude position. It had a maximum scan rate of 20 degrees per minute over a range of +/-9 degrees, which allowed it to track the shuttle, a satellite in low earth orbit, or an aircraft. The SBA feed, which was connected to the west hopping beam family, could be used to track moving objects by physically rotating the antenna reflector. The antenna employed an

20 GHz transmit feed assemblies containing the beam-forming networks. The right assembly is the east family hopping beam. EMS Technologies produced the fast, low loss, beam-forming switch networks.

offset-fed, parabolic, graphite fiber-reinforced composite reflector that was 40 inches in diameter. The reflector feed was fixed and consisted of a diplexed, corrugated, dual-band horn. Due to its smaller size, the gain of the SBA is approximately 6 dB less at 20 GHz than that of the MBA. Pointing is accomplished in two-axes by gimbal drives, which rotate the reflector surface in pitch and roll using open-loop control. With a beam width of approximately 1.0 degree at 20 GHz and 0.75 degree at 30 GHz, the ground coverage spot is approximately 400 miles in diameter. Dupont Chemical manufactured the SBA using the same materials and fabrication techniques as the main reflectors. Special software developed for the ACTS program to evaluate the com-

43

plex contours of the antennas proved invaluable. Using this software, errors were uncovered in the mold for the SBA. These errors were corrected before reflector fabrication began. Without this software, the errors would not have been detected until RF range testing was conducted much later in the development cycle. Therefore, the software tool prevented costly re-fabrication and schedule delays.

Spacecraft in the folded launch configuration, with thermal insulating blankets (gold) installed. Mechanically steered antenna is located directly above black solar panel. Left to right: Frank Gargione, systems manager; Paula McGrath; and Mike Kavka, program manager for Lockheed Martin Astro Space, who was prime contractor for the spacecraft.

Antenna Support Assembly The antenna reflectors and feed assemblies were mounted to a low-distortion truss system that was attached to the spacecraft's Earth-facing panel. The truss system was built with graphite epoxy tubes and titanium end fittings. The entire antenna assembly weighed about 900 pounds, with the truss making up about half the weight.

Antenna Pointing Once ACTS achieved orbit and deployed its antennas, the transmit and receive spot beams were brought into co-alignment. This was accomplished by slightly rotating the transmit main reflector, using its two-axis drives. These in-orbit fine adjustments were necessary to remove beam mispointing due to residual gravity stress, assembly stress, alignment errors, and hydroscopic reflector distortions. The narrow beam width and high gain slope of the MBA required a pointing accuracy that was 10 times better than conventional CONUS-coverage antennas—i.e., of the order of 0.025 degrees maximum in pitch and roll and 0.15 degrees in yaw.

To achieve this pointing accuracy, an autotrack system based on Landsat technology was used to determine pointing errors in pitch and roll [34]. The autotrack error signals are referenced to the continuously maintained command carrier, uplinked from the master ground station in Cleveland, and received by the Cleveland fixed horn. Note that only the received reflector system was autotracked. Yaw attitude is determined twice a day using sun sensors located on the east and west faces of the spacecraft. A software estimator provides the control signal during the periods when the sensors are not viewing the sun.

A momentum wheel and two magnetic torquers are used, respectively, for correcting pitch, roll, and yaw errors. In-orbit test data demonstrated that the attitude control system provides the required pointing accuracy for the MBA.

Development Approach It was considered prohibitively expensive to test the RF performance of the MBA under simulated, in-orbit thermal conditions. The approach taken from the very beginning of development was to rely on thermal and RF modeling to verify that the MBA was suitable for flight operations. As discussed below in the in-orbit performance section, this higher-risk but lower-cost development approach was not 100% successful. As it turned out, accurate modeling of the complex MBA was not completely accomplished even though a separate NASA design verification team—using detailed, independent models—checked the contractor design.

Assembly & Test A major challenge in the assembly of the MBA was to ensure that each component of the transmit and receive antenna was mechanically aligned (with respect to its assembly and the other antenna) such that the cor-

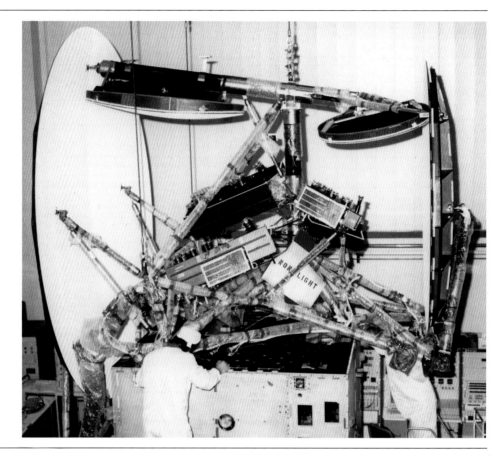

Antenna assembly being fastened to spacecraft bus at three attach points. Clearly shown is the antenna support truss holding the main reflectors, subreflectors, and feed assemblies. Wrapping on support tubes is for protection and is removed before launch.

responding spot beam patterns would coincide with an accuracy of 0.015 degrees. In addition, it was necessary to continually check this alignment as the spacecraft went through various environmental tests. The reflectors, subreflectors, feed assemblies, and the antenna support structure were aligned using a theodolite-based optical technique. To characterize the RF patterns and gain characteristics of the MBA, a spherical near field antenna test facility was constructed. The MBA was first tested on a spacecraft mockup, and finally on the flight spacecraft itself. Tests were conducted before and after each environmental exposure to confirm that no changes had occurred to the antenna alignment.

Interface Between the MBA and the Communication Payload Uplink signals can be received through the five antenna ports. Each of the uplink signals can be directed to any of the four receivers by means of the waveguide input redundancy switch. Two beams can be connected to the BBP, or any three beams can be connected to the MSM inputs. The four low-noise receivers amplify the 30 GHz communication signals and down-convert them to 3 GHz for routing by either the BBP or MSM. The receiver outputs can be directed to either the BBP or MSM using the receive-coaxial-switch-assembly. The receivers provide linear outputs to the BBP and saturated outputs to the MSM. After processing by either the BBP or MSM, signals are switched through the transmit coaxial switch assembly to any of the four upconverters. The upconverters convert the IF signals to 20 GHz for amplification in the 46-watt traveling wave tube amplifiers (TWTA). The outputs of the TWTA can be directed by the waveguide output redundancy switch to any of the three fixed beams or the two hopping beam ports of the transmit antenna.

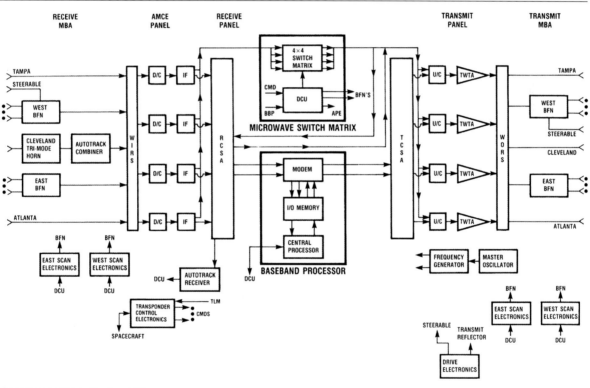

Communications Payload Diagram

Baseband Processor (BBP)

The ACTS BBP routes individual voice, video, data, and multimedia messages between a large number of small terminals on demand [35,36]. Key advantages of the ACTS BBP are the capability to:

1. automatically reconfigure message routing to accommodate dynamic traffic changes without the use of a terrestrial hub and double-hop transmissions;

2. provide additional link performance by the decoupling of the RF uplinks and downlinks;

3. selectively apply forward error correction coding and burst rate reduction on individual messages to overcome localized rain fade; and

4. allow different uplink and downlink multiple access and data rates.

The BBP operates with two uplink and two downlink hopping beams. Each one of the four is time-independent to provide flexible operation. All of the user terminals in the two hopping beam families (east and west) are completely interconnected in a mesh network through the satellite switch. The BBP operates in a TDMA format with a 1 millisecond time frame. Serial minimum shift key (SMSK) modulation is used for both the up- and downlinks, and was chosen because of its lower side-lobe levels. A unique feature of the BBP is that the uplink and downlink access formats are different. Uplink bursts at either 27.5 or 110 Mbps use combined Frequency Division Multiple Access (FDMA) and TDMA access. Downlink bursts at 110 Mbps are transmitted in a single channel per beam in TDMA mode.

After an uplink burst arrives at the satellite, it is demodulated and stored in input memories. During the next frame, the data is read out of the input memories, routed through a digital switch, and stored in output memories for the proper downlink beam. During the third frame, the output memories are read out and the data transmitted. The amount of BBP memory is largely dependent on the TDMA frame period. The shorter the frame period, the smaller the memory size required, resulting in less weight and power. TDMA frame efficiency (ratio of message bits to overhead bits) is higher for a longer frame period. An ACTS spacecraft power and weight versus frame efficiency tradeoff resulted in a selection of a 1 ms frame period for both the uplink and downlink.

The BBP switches on a word-by-word basis, where a word is equivalent to a 64 Kbps message channel. Any mix of voice, video, data, or multimedia messages in increments of 64 Kbps can be accommodated. The four-foot user terminal can handle 28 64 Kbps channels for a maximum information throughput of 1.79 Mbps.

Demand Assignment and the Orderwire Concept All active terminals in the network are assigned dedicated two-way control channels, called *orderwires*, through which they maintain communications with the Cleveland master control station (MCS). In the BBP network, the MCS is the centralized network control and monitoring facility. All traffic requests and circuit assignments are processed by the MCS. Users can change their requests for circuit capacity once every superframe—which is 75 ms. The BBP operates under stored program control. The control memories direct the operations of demodulators, decoders, the base band routing switch, encoders and modulators, and the hopping sequence of the MBA's spot beams. In response to traffic requests, the control memories are programmed and dynamically modified by the MCS, which is providing demand-assigned multiple access (DAMA) operations.

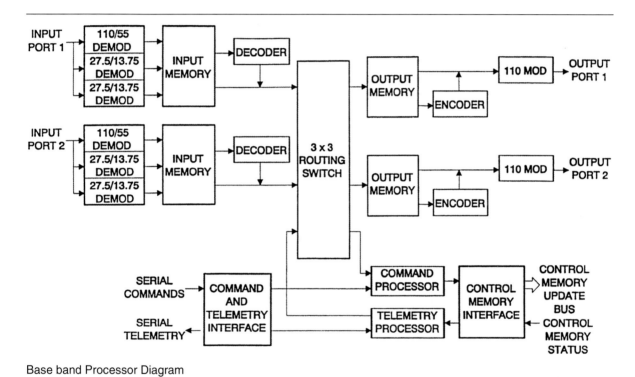

Base band Processor Diagram

Two control memories are used in the BBP—one on-line and one off-line. The on-line memory contains the current beam-hopping and switch-routing instructions. The off-line memory contains the new beam-hopping and switch-routing instructions and is held in readiness for execution to on-line status at the instant of reconfiguration. When it is time to reconfigure the BBP to a new routing program, the control memories are swapped. This swap is performed in synchronization with changes in the burst time plans for terminals throughout the entire network. Ongoing traffic is not interrupted.

Rain Compensation A key feature of the BBP is its capability to automatically implement coding and burst rate reduction to only those TDMA bursts experiencing fading conditions. During periods of rain, the RF transmitted and received signals can be significantly attenuated at Ka-band frequencies. When the downlink signal attenuation at the earth stations exceeds a threshold, rain compensation is automatically applied. This consists of reducing the burst rate by a factor of two on the terminal uplink and downlink, and simultaneously incorporating forward error correction (FEC) coding. This increases by a factor of four the number of time slots allocated to the station. In order to avoid affecting all the other stations' burst times, the burst time for the affected station is moved from the clear part of the frame to a fade pool area reserved for providing additional time slots to stations requesting compensation. If fixed margins were utilized instead of adaptive compensation, the effective communications system throughput would be reduced by a factor of four. The amount of capacity reserved for fade compensation is variable and can be adjusted as needed by the specific rain area of operation. The FEC process employs a maximum likelihood convolutional code having a rate of 1/2, a constraint length of 5, 2-bit soft decision quantization, and path memory length equal to 28. This combination of burst rate reduction along with FEC coding provides an additional 10 dB improvement in link margin. Fixed margins of 3 dB on the downlink and 5 dB on the uplink were provided to accommodate small fades without compensation. These fixed margins plus the 10 dB gain provided by rain compensation nominally allow the bit error rate to be maintained at 5×10^{-7} with service availability for CONUS of 99.5%.

This rain compensation is adaptive and only implemented when the fade for the 20 GHz signal exceeds a threshold. Using the in-band communication channel, each VSAT continuously estimates its downlink signal level. When the fade threshold is exceeded, the compensation is automatically applied within 1 second. When the fade magnitude reduces to a predefined cessation level, the rain compensation is removed. Since a very small percentage of user terminals simultaneously undergo fades exceeding the threshold at any time, only a small amount of onboard decoding capacity is needed. In the case of ACTS, the amount of decoding capacity is 6.8 Mbps per beam.

Burst Timing In order to limit guard time between bursts, it was decided that an uplink burst arrival at the BBP must be accurate to within +/-60 nanoseconds (ns). The BBP uses a correlator to detect a 7-bit unique word and determines whether it is early or late. This information is transmitted to the user terminals in the form of a tracking error word (TEW). The ground terminals then adjust their uplink transmission bursts in the next frame according to the

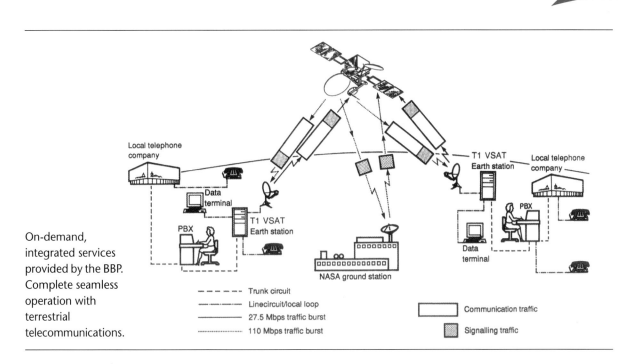

Local telephone company

Data terminal

PBX

T1 VSAT Earth station

NASA ground station

T1 VSAT Earth station

Local telephone company

PBX

Data terminal

On-demand, integrated services provided by the BBP. Complete seamless operation with terrestrial telecommunications.

- - - - - Trunk circuit
—·—·— Linecircuit/local loop
———— 27.5 Mbps traffic burst
············ 110 Mbps traffic burst

Communication traffic

Signalling traffic

message contained in the TEW in order to maintain time synchronization with the spacecraft.

BBP Development Motorola developed the architecture for the BBP over a number of years, starting in 1979 [35]. To prove that the BBP could be produced with an acceptable weight and power, a proof-of-concept (POC) model was developed from 1980 to mid-1983. As part of this NASA POC program, Motorola developed custom, large-scale integrated (LSI) circuits for certain high-speed functions of the demodulators, control memory update controllers, and decoders. This pre-flight POC program development successfully proved the feasibility of the switchboard-in-the-sky.

COMSAT Laboratories led the development of ground/spacecraft architecture for the TDMA, DAMA, system synchronization, rain compensation, and network control. This was a very large and complex task that involved a high degree of coordination between NASA (the system integrators) and the spacecraft contractors. COMSAT's contribution of the BBP architecture was as important as Motorola's contribution of hardware. Both Motorola and COMSAT Laboratories are to be commended for their principal roles in the development of such a revolutionary technology.

BBP Hardware Implementation The BBP hardware was partitioned into three flight assemblies: a modem unit, an input/output memory unit, and a central processor unit [37]. Hardware partitioning was based on the common design requirements for elements of the boxes and the optimum allocation of signal and data interfaces. The modem unit housed six uplink demodulators and two downlink modulators. The input/output unit contained the input and output channel data memories, the decoders and encoders, the data routing switch, and the input and output channel control memories. The central processor received commands from the MCS and updated all control memories. The flight BBP weighs 121 pounds and consumes 199 watts of power.

The low-power, high-speed processing requirements, coupled with the space environment, posed significant challenges to the BBP design. The key technologies developed by Motorola to enable the ACTS BBP included:

1. compact, low-power rapid acquisition burst demodulators operating at 27.5 Mbps to 110 Mbps;

2. forward error correction utilizing a single chip maximum likelihood convolution decoder;

3. a central processor architecture that provides circuit-switching via commands from the master control station; and

4. a family of high-speed, low-power, large-scale integration (LSI) and small-scale integration (SSI) circuits.

The expected two-year radiation dose was 10 KRAD. The LSIs, SSIs, and memories were tested at much higher levels. The LSIs and other integrated circuits were also life-tested for 1,000 hours at 150°C junction temperature.

Microwave Switch Matrix (MSM)

The ACTS MSM is an IF switch capable of routing high-volume point-to-point and point-to-multipoint traffic. It is a solid state, programmable, 4x4, planar *cross-bar* switch [38]. Functionally, only a 3x3 configuration is used for connecting any three uplink beams to any three downlink beams. The 4x4 hardware implementation provides a three out of four redundancy. Unlike the BBP, the MSM does not store the incoming traffic messages in onboard memory. The cross points of the matrix switch use two single-gate, field effect transistors (FET) in a hybrid switch/amplifier configuration. It was necessary to configure two FETS in series to meet the channel isolation requirements. The MSM routes signals over a 900 MHz-wide transmission channel (1 dB bandwidth, amplitude limited) between 3 to 4 GHz. Input signal distribution and output signal combining are performed using passive recursive couplers. The outputs from the IF modules are run at saturated levels to ensure that the inputs to the MSM are all equal, thereby minimizing the cross talk between MSM signal paths. The very fast (less than 100 nanoseconds) switching time

Three in-line BBP boxes built by Motorola. Front to back: the central processor, the input/output and modem boxes mounted on the spacecraft's north panel. To the left of the central processor is the microwave switch matrix.

of the Gallium Arsenide FET amplifier switches permits efficient dynamic routing for use with TDMA communication traffic.

The simple connect-disconnect feature of the switch matrix allows for a variety of uplink-to-downlink antenna connections. A selection of beam-to-beam or broadcast connections that join a single uplink beam with several downlink beams is permitted. The waveguide input redundancy switch (WIRS) permits the selection of any three uplink beams from the five MBA antenna ports. Likewise, a waveguide output redundancy switch (WORS) allows the selection of any three downlink beams. These switches are electro-mechanical and are generally fixed for long periods of time.

A programmable, digital controller with dual memory banks controls the MSM. This digital controller, commanded directly by the 5 Kbps spacecraft command channel, stores the matrix switch and antenna-hopping sequence data for a TDMA frame. For each increment of time during a TDMA frame, the controller sequentially reads out the memory and changes the matrix switch connectivity and hopping beam positions. In this manner, the TDMA bursts from terminals are sequentially switched on board the satellite to the proper downlink beams. The present TDMA plan is stored in the foreground memory. The background memory is programmed through the spacecraft command link to implement the next TDMA plan. A memory swap command from the ground is issued to change to a new TDMA burst time plan when required by the communication traffic.

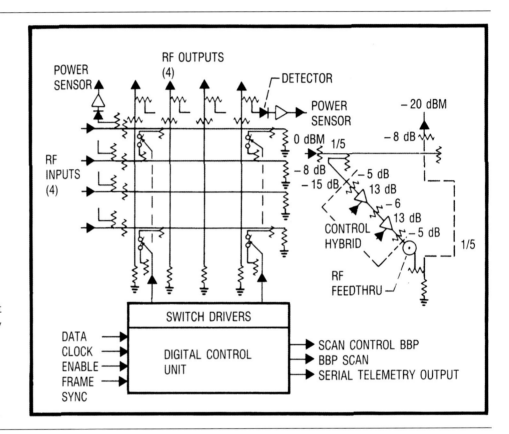

Microwave switch matrix operation at 3 GHz IF frequency with switching speed less than 1 microsecond.

Built by Lockheed Martin Astro Space.

RF Ka-band Components

In addition to the MBA, BBP, and MSM, RF Ka-band components were identified to be the high-risk technologies needed to support future commercial Ka-band satellites. The ACTS developments for 20 GHz transmitters and 30 GHz receivers are described below.

20 GHz High Power Transmitter The ACTS transmitters consist of solid-state frequency upconverters followed by traveling wave tube amplifiers (TWTA) and provide operation over a 900 MHz pass band. Each TWTA used in the ACTS spacecraft consisted of a separately packaged 46-watt, traveling wave tube (TWT) designed and manufactured by Watkins-Johnson (which was subsequently purchased by Varian and then Hughes), and a high voltage electronic power conditioner (HV EPC) manufactured by Lockheed Martin.

The ACTS TWT was originally designed for dual-mode operation and achieved a DC to RF conversion efficiency of 43% and a low phase distortion (4°/ dB AM-PM). This design was an outgrowth of Watkins-Johnson Ka-band flight tube developments for both the European Space Agency's Olympus program and the U.S. Navy's FLT SAT COMM program. The tube operated in the saturated mode with a gain of 52.5 dB, and was completely stable against oscillations. The saturated mode was chosen because the primary mode of communication on the downlink was single carrier TDMA and not multi-carrier operation. By careful mechanical design and the use of low-outgassing potting techniques, Watkins-Johnson developed a very compact, lightweight tube that weighed only 2.8 pounds. TWT and power conditioner technology has improved today to provide much higher efficiencies.

30 GHz Low Noise Receiver The ACTS receivers developed by Lockheed Martin amplify the 30 GHz, TDMA uplink signals from the selected MBA beams and down-convert them to the 3-4 GHz IF signal for routing purposes. To achieve the required low-noise, high-gain performance, the first three amplifying stages of the receiver employed high-electron mobility transistors (HEMT).

Four 20 GHz TWTs (left) and upconverters (right) mounted on the spacecraft's south panel.

Upconverters manufactured by Lockheed Martin and TWTs by Watkins-Johnson.

HEMT devices have similarities with conventional GaAs FETs but differ in their carrier transport mechanism. HEMTs are fabricated on an AlGaAs/GaAs heterostructure and use metal electrodes. These devices have demonstrated lower noise and higher gains at 30 GHz than conventional GaAs devices. The HEMTs used in the ACTS receivers were manufactured by Lockheed Martin Electronics Laboratory (previously GE) in Syracuse, New York, with an individual noise figure of 1 dB and a minimum gain of 11 dB. The three-stage amplifier had a rated noise figure of 3.4 dB beginning-of-life, 3.6 dB end-of-life, and an overall gain of 20 dB. The large bandwidths (900 MHz) and high frequencies (30 GHz) required the development of new methodology in manufacturing, alignment, and testing. The performance of the hardware was much more sensitive to lead lengths and grounding than at the lower frequencies. New techniques were developed to provide consistent grounds, low voltage standing wave ratio connections, and adequate stress relief. Alignment required microscopic monitoring to determine tuning locations. Extensive use of test fixtures was employed to provide accurate and repeatable data.

Spacecraft Bus

The spacecraft bus was based on the Lockheed Martin (RCA) 4000 Commercial Series. The bus structure was a rectangular box, roughly 70 inches along the vertical axis with a cylindrical center structure that housed the apogee kick motor (AKM). The antenna support assembly for the MBA was attached to the bus-structure at only three points to limit thermal distortion in the MBA. The north- and south-facing panels were each divided into three panels. These panels were used to mount most of the spacecraft bus and communication payload electronics equipment. The LNRs and BFNs were, however, attached to the MBA support structure to limit the length of the 30/20 GHz waveguide runs.

The electrical power subsystem is a direct energy transfer configuration consisting of solar array panels, storage batteries, and power regulation equipment. The four solar array panels (135 sq. ft.) provided a beginning-of-life power of 1836 watts.

The propulsion system consisted of a hydrazine reaction control subsystem and the apogee kick motor. The catalytic thrusters, propellant tanks, and the plumbing comprised the reaction control subsystem. This was configured as a blow-down system utilizing hydrazine propellant pressurized with helium. Sixteen catalytic thrusters, ranging from 0.2 to 1.0 pound of thrust, are used primarily for orbit adjustment maneuvers but also participate in certain attitude control functions. Five hundred and eighty pounds of hydrazine propellant were on board at liftoff. The apogee kick motor was the Thiokol solid-propellant STAR-37 FM model that was fired to move the satellite from the elliptical transfer orbit to the circular geostationary orbit.

Thermal control used a combination of heaters, a selection of finishes, and multi-layer insulation blankets to maintain proper temperatures. Heat pipes were used under the traveling wave tubes and the baseband processor to remove heat efficiently from these high heat density components and thus limit the upper extreme of the temperature range.

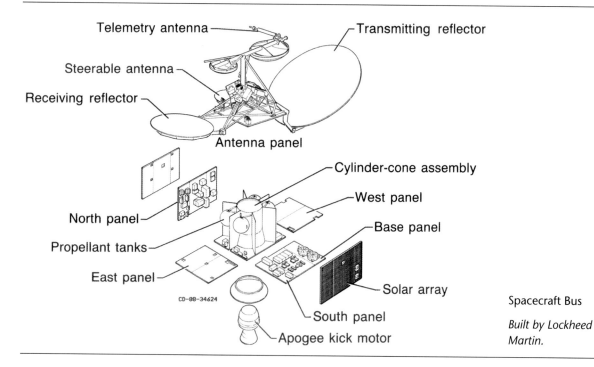

Spacecraft Bus

Built by Lockheed Martin.

For attitude control, a momentum wheel controls the spacecraft pitch attitude and a magnetic torque coil controls the roll attitude using error signals from either the autotrack or the earth sensors. A magnetic torque coil also controls the yaw attitude using signals from two sun sensors or a yaw estimator.

In the earth sensor mode, the design requirement for roll, pitch, and yaw accuracy was 0.10, 0.10, and 0.25 degrees, respectively. In the autotrack mode, the design requirement for roll, pitch, and yaw accuracy was 0.025, 0.025, and 0.15 degrees, respectively [34]. The earth sensor mode of attitude control is primarily used during the eclipse shutdown period, when the communication payload is turned off. The earth sensor is also used daily during the thermal transients of the subreflectors, which affect the autotrack accuracy.

The ACTS flight system incorporated four Ka beacons for real-time fade measurement [28]. Two of these beacons are in the downlink frequency band

Lockheed Martin spacecraft bus. Two of four hydrazine tanks mounted facing the east panel. On the left, four batteries are mounted on the north panel. Solar panel drive shaft is shown on the extreme right.

while the others are in the uplink band. These beacons provide signal sources to make continuous measurements for propagation research. The downlink frequency beacons operate at 20.185 GHz with vertical polarization or at 20.195 GHz with horizontal polarization. These beacons primarily provide the normal spacecraft telemetry and ranging functions while producing a stable downlink signal to allow propagation measurements.

Both uplink frequency beacons operate at 27.505 GHz. Each beacon is vertically polarized and not modulated. In contrast with the telemetry beacons, which can operate simultaneously, only one uplink beacon can be powered at any one time. The 27.505 GHz frequency was selected to avoid interference with the communication signals. The beacon signals were derived from independent local oscillators and, therefore, were not coherent to each other. The beacon antennas provide broad coverage—primarily to the continental United States.

In-Orbit Performance

ACTS was launched on the space shuttle Discovery (STS-51) on the morning of September 12, 1993. After deployment, the burn of Orbital Sciences' solid rocket stage resulted in placing the spacecraft in a transfer orbit that was within one sigma accuracy. This excellent performance of Orbital's TOS stage maximized the amount of hydrazine for in-orbit operations. A lot of the credit for the new TOS vehicle performance went to the NASA Marshall Space Flight Center at Huntsville, Alabama, which managed its development. The spacecraft arrived at its permanent orbit position at 100 degrees west longitude on September 28, 1993. At this time, ACTS entered into its pre-operational mission phase, which consisted of spacecraft checkout followed by system and communication network checkouts. With the completion of all spacecraft, ground system, and network checkouts, the operations program was initiated on December 1, 1993. ACTS in-orbit tests showed that the communications payload and bus performances were in close agreement with the ground test results conducted during spacecraft development and assembly [39, 40, 41 and 42]. After over six years of in-orbit operations, ACTS has exceeded its two-year lifetime requirement has proved to be extremely reliable, and has lost none of its communication capability.

Baseband Processor (BBP)

The demand-assigned processing and routing of individual voice, data, video, and multimedia messages between multiple earth stations, via the hopping beams, was routinely accomplished by the BBP Pre-launch reliability models showed the BBP as the overall driver of communication payload reliability due to its circuit complexity and high parts count.

The payload probability of success was calculated as 0.78 at a 36% duty cycle for two years for 220 Mbps information throughput. No failures have occurred in the BBP, except for a single anomaly attributed to a control memory bit latch-up that was cleared by a power recycle.

Testing and experimentation have shown that the spacecraft and the T1 very small aperture terminals (VSAT) successfully met design performance

59

specifications and requirements for acquisition, synchronization, timing, and message routing. The acquisition, synchronization, and timing process remained functional at 30 GHz in rain fades of up to 15 dB. Bit error rate (BER) performance ranged from better than 1×10^{-11} in clear sky conditions to not less than 5×10^{-7} in 15 dB rain fades when rain fade compensation was enabled.

The STS-51 space shuttle crew which deployed ACTS. Left to right: William Readdy, Dan Bursch, Commander Frank Culbertson, Carl Walz, and Jim Newman.

Data has shown that the ACTS adaptive rain fade compensation successfully provides enhanced link margin automatically as needed [43, 44, 45]. The transitions from uncoded to coded operation and back were accomplished with no loss in throughput and without errors.

The ACTS T1 VSAT employs a terrestrial interface unit that consists of a small programmable central office. This interface supports a variety of Bell standard hardware interfaces to provide seamless interconnectivity into the terrestrial telephone network—an important requirement for satellites operating in future national and global networks (NII/GII). The interface control software has been custom-designed to provide protocol conversion between terrestrial circuit connect and disconnect protocols and the VSAT DAMA protocols. Analog voice is encoded by circuit line cards in the terrestrial interface at a rate of 64 Kbps. In addition, an echo-canceling device is incorporated into each interface circuit card. This device employs digital signal processing and does an excellent job of eliminating all noticeable echoes. ACTS users found, under these conditions, that "the propagation delay was unnoticeable" for voice calls. Users of ACTS frequently say they do not believe they are talking over the satellite. The fact that ACTS users have found the satellite delay unnoticeable seems to be contrary to previous experiences for geostationary satellites. We believe the reason for this is that in the past the echo was not consistently eliminated under all conditions and that voice quality was deficient. Extensive laboratory tests have been conducted and they support the conclusion that *echo*—not delay—is the principal cause of dissatisfaction to users.

The ACTS VSAT user interface also supports the connection of an Integrated Services Digital Network (ISDN) Primary Rate Data Interface (PRI). Operating at 1.544 Mbps, the PRI provides 23 communication channels and 1 signaling channel. Using this satellite ISDN capability, desktop conferencing quality was equal to that provided by terrestrial ISDN, call set-up times were fast (taking only 2 to 3 seconds), and multi-point video conferencing was reliably handled with the development of special interface software. Overall, seamless, high-quality, satellite/terrestrial ISDN services were reliably provided. The general conclusion from the ACTS ISDN program is that basic and primary rate ISDN can be readily incorporated into future satellite systems.

Microwave Switch Matrix (MSM)

The MSM mode of operation allows the full 900 MHz bandwidth of the transponders to be used with the hopping or fixed beams. Since launch, the Link Evaluation Terminal (LET) located at NASA LeRC has been periodically used to check out the signal paths through the transponder and the MSM. Close agreement has been obtained between the expected and measured values. TWTA output levels have remained steady at approximately 46 watts, providing a peak EIRP of 58-69 dBW depending on the downlink beams. TWTA parameters, monitored via telemetry, have remained stable with over 53,000 hours (as of December 1999) on each of the three tubes normally used. In addition, no spurious shutdowns have occurred. A continuous wave tone, uplinked from LET, swept across the communication band, has shown close agreement with ground test data across the 900 MHz bandwidth.

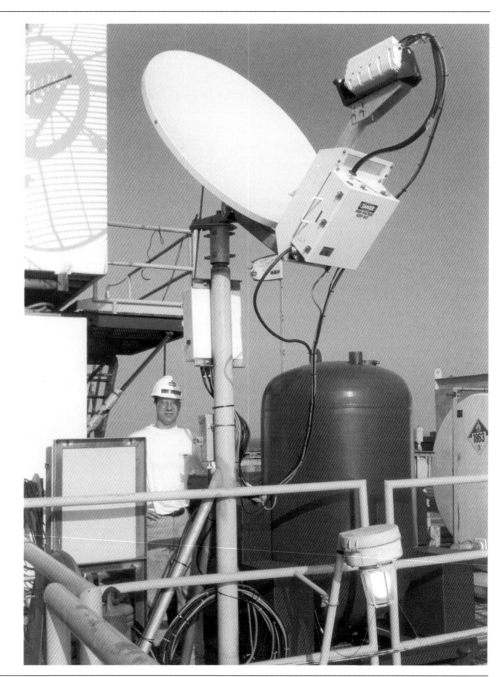

Amoco used ACTS T1-VSATs to transmit real-time data from its offshore oil platform in the Gulf of Mexico to its onshore processing center. With the excellent BER performance of ACTS, data was transmitted using ATM.

Gigabit Network

The ACTS gigabit network provided point-to-point and point-to-multipoint full duplex services using the satellite-switched TDMA capability of the MSM, along with the hopping-beam network [46]. The network and the associated high-data-rate (HDR) ground terminals were developed by GTE (formerly BBN Systems and Technologies) and Motorola under joint NASA and Defense Advanced Research Project Agency (DARPA) sponsorship. The user interfaces were compatible with Synchronous Optical Network (SONET) standards and were readily integrated with standard SONET fiber-based terrestrial networks. The two rates supported were OC-3 (155.54 Mbps) and OC-12 (622.08 Mbps). In addition, asynchronous transfer mode (ATM) communications could be run readily on top of the SONET structure.

The network control and management functions were contained in each HDR earth terminal with the operator's interface being centralized in a portable network management terminal (NMT). The NMT could be located at any HDR earth terminal site or, alternatively, at any location with a terrestrial Internet connection to any HDR earth station designated as the reference station.

Transmissions to the satellite were performed at 348 Mbps or 696 Mbps with staggered or offset BPSK (SBPSK) and staggered QPSK (SQPSK) modulations, respectively. Using the 11.5-foot user terminals, the network provided an availability of 99% within CONUS for transmissions at 696 Mbps. All transmissions used a 232,216 Reed-Solomon block error correction code to achieve a BER of 10^{-11}.

ACTS tests and experiments have demonstrated the physical layer compatibility of the satellite SONET implementation with terrestrial SONET equipment and networking. In addition, asynchronous transfer mode (ATM) services have been implemented using the SONET structure and have demonstrated seamless satellite/terrestrial ATM with low BER ($<10^{-11}$).

Multi-beam Antenna (MBA)

A number of MBA performance evaluations have been conducted since launch [47]. These tests were designed to evaluate beam-pointing stability, beam shape, antenna gain, sidelobes, cross polarization, and so forth in the space environment. The test measurements found the MBA performance to be well within the design and pre-launch test range, with the exception of beam pointing. Thermal effects created greater-than-expected pointing errors for the 0.30 degree spot beams. These in-orbit thermal distortion effects on the MBA can be classified as:

- rapidly varying, or

- diurnally varying

The rapidly varying thermal distortion is caused by non-uniform sun illumination (the sun/shadow line moves across the face of the reflector) on the front surface of both transmit and receive subreflectors. This thermal distor-

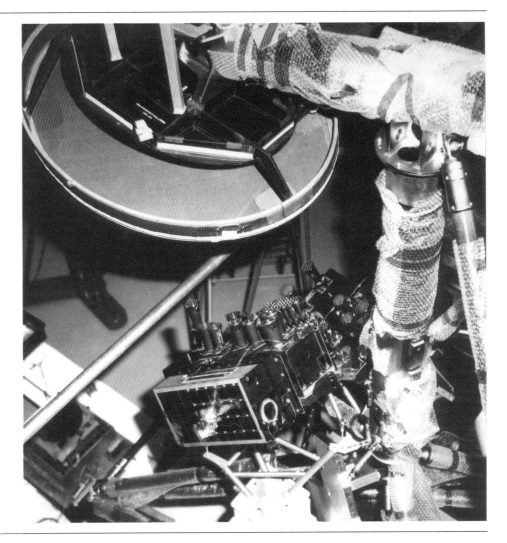

The MBA subreflector assembly. Front surface with outer ring is depicted by the orange material. The rear surface is black with rear egg crate support. Feed housings are shown below the subreflector.

tion causes a large temperature gradient in the front face of the subreflectors, resulting in surface distortion on the order of 0.060 inches peak. This, in turn, produces beam movement or wandering of approximately 0.1 degree. The large subreflector temperature gradients occur for a total of approximately two hours each day and correlate with the rapidly varying beam wandering. Because this large thermal distortion occurs just on the front face of the subreflector, only the east beams are affected.

The diurnally varying thermal distortion causes a westward movement of the transmit beams starting at approximately 0200Z–0400Z each day, reaching its maximum of approximately 0.2 degrees at 0800Z and returning to the starting point by 1400Z. It is believed that this variation is caused by thermal expansion of the spacecraft bus, which results in an apparent rotation in pitch of the transmit main reflector with respect to the receive reflector which is held fixed to Cleveland by the Autotrack signal. Using biaxal drive gimbals, the transmit reflector is adjusted in pitch twice daily to compensate for this diurnal thermal movement.

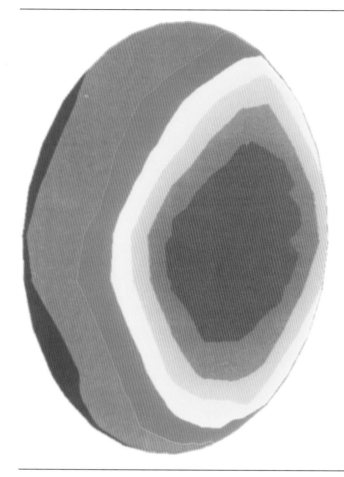

Post-flight thermal gradient predictions for the front face of the subreflector assembly. From dark blue to deep red the thermal gradient is 221 degrees Fahrenheit.

In addition to these thermal distortions, two MBA non-thermal distortions have been observed. A 1 Hz oscillation distortion—affecting all downlink beams—is caused by induced mechanical noise on the transmit main reflector. The mechanical noise is generated by tiny step changes in wheel speed of the momentum wheel, which cause movement in the transmit reflector around the biaxial drive pivot. The oscillation causes minimum signal variation (<0.1 dB) at beam center that becomes large (>3 dB) at beam edge. The amplitude was measured peak-to-peak to be 0.075 degrees. In addition, yaw errors are introduced by the attitude control system.

Using daily attitude control biasing, the beam-pointing errors caused by thermal distortions have been minimized on ACTS so that acceptable communication performance could be achieved throughout the day. The ACTS beam-pointing problems do not represent a technology barrier, but do point out the need for careful thermal and mechanical design for future MBA systems that have narrow spot beams. Since yaw errors can also cause significant MBA signal variations, future spacecraft need to improve the capability for more precise yaw control.

Spacecraft Pointing Accuracy

In addition to thermal stability (see the above section), the MBA pointing performance relies on precise spacecraft pointing. The spacecraft-pointing accuracy achieved in roll, pitch, and yaw was 0.025, 0.025, and 0.20 degrees or less, respectively. Only the yaw accuracy was slightly greater than its design requirement (0.15 degrees). This 0.05 degree greater yaw error, however, does not have a significant impact on the MBA performance.

Beacons

The uplink fade beacon at 27.5 GHz provides an unmodulated uplink band signal to propagationists. The primary uplink fade beacon unit was found to be 4 dB low in RF power output when the system was turned on after launch. The backup unit was turned on in November 1993 and satisfactorily operated. In December of 1999, the primary uplink beacon was turned back on and was found to be operating at the proper power level. Since both beacons are connected to the antenna through a mechanical waveguide switch, the power shortfall is attributed to a mechanical misalignment of the switch rotor caused by launch vibrations. This has been the only redundant unit brought into service because of a hardware malfunction since launch. The frequency of both the uplink and downlink Ka-band beacons continues to be very stable, measured as <1800 Hz diurnal variation and no more than –7400 Hz yearly drift.

The only other RF anomaly to occur was with the C-band beacon. In April 1999, during a periodic rehearsal of the backup C-band station, one of the two beacons was found to be 10 dB low. No reason for this anomaly has been found.

System Reliability

The ACTS spacecraft payload and subsystems have accumulated over six years of in-orbit operations. Other than the switch to the redundant uplink beacon shortly after launch, as previously mentioned, all other systems and subsystems functioned with their primary units. The new switching and processing advanced technology units performed well after year-round, 24-hour/day operations. The MBA while providing satisfactory RF links, has revealed the need for careful thermal design to prevent spot beam motion.

The overall success of ACTS is attributed to extensive design analysis, performed by the NASA-industry teams and a rigorous ground test program. Future system operators of ACTS-type commercial satellites should feel confident that they can put their systems into service with acceptable risk.

CHAPTER 3

TERMINAL EQUIPMENT

The History of Satellite Terminals

One of the important advantages of Ka-band, spot beam satellites is the capability to use very small, low-powered, low-cost, and high-data-rate user terminals. These terminals represent a paradigm shift for the application of communication satellites. An appreciation of this shift can be gained by tracing the history of satellite terminals.

The first commercial communication satellite application occurred in 1965, with the launch of Early Bird (INTELSAT 1) for international service. The standard antennas for this service were nearly 100 feet in diameter and used cryogenically cooled receivers. Only large common carriers could afford to build and operate these stations.

Since 1965, there has been a steady development of satellites with increased transmitter power and higher-gain antennas covering smaller geographical regions—this allows the use of smaller-diameter terminals that are lower in cost. The early 1970s saw the emergence of domestic satellite services for Canada and the United States. These satellites—which had medium-powered transmitters and concentrated that power on a smaller region (e.g., Canada or the United States)—permitted the use of small (16-32 feet) and medium-sized (49 feet) earth stations. All of these satellites operated at C-band (uplink at 6 GHz and downlink at 4 GHz).

In 1976, NASA and Canada launched CTS-1—a high-powered Ku-band satellite (200 watt transmitter) designed to operate with small earth stations. CTS-1 proved the feasibility of service at Ku-band frequency (uplink at 14 GHz and downlink at 12 GHz) for video reception by small transportable terminals less than 3 feet in diameter. In 1978, COMSTAR 3 and ANIK B were the first two Ku-band satellites to provide domestic services in North America. At that time, American Satellite Corporation introduced one of the first business services when it provided 56 Kbps circuits using 15- and 32-foot terminals [48]. In 1980, Satellite Business System (SBS) was launched to provide business voice, video, and data services between 18-foot terminals located on the customer's premises. SBS, whose partners included COMSAT and IBM, was specifically targeted for business telecommunications. The terminals were too large and expensive, however, to effectively compete with terrestrial communications.

Very Small Aperture Terminals (VSAT)

The mid-1980s saw higher-powered satellites (e.g., SATCOM and GSTAR series), which permitted still narrower antenna beams (half of the United States). This combination of more radio frequency power in the satellite and a more concentrated beam permitted the use of still smaller terminal antennas,

Large C-band
Antenna Required
for Communication
Services in the
1960s

which cost much less. These satellites permitted 4- to 8-foot terminals to pro-vide low-data-rate services (up to 56 Kbps) into a large hub station. In order to provide interconnectivity between two user terminals, undesirable double-hop communication was necessary. Today, this size class of terminal for voice and data services is commonly called a very small aperture terminal (VSAT).

Since the mid-1980s, VSAT services have become one of the prime uses of satellite communication with over 360,000 star TDMA terminals installed worldwide as of August 1999 [49]. The satellite powers and antenna gains (e.g., ASIASAT 3) continue to increase with the Ku-band Effective Isotropic Radiated Power (EIRP) and Gain/Noise Temperature (G/T) reaching a maxi-mum of 55 dBW and 7 dB/K, respectively. Such satellites allow 4- to 8-foot VSATs to communicate with one another in a single-hop, 256 Kbps TDMA burst mode. The costs for 4-foot VSATs for several voice circuits are on the order of $3,000 to $5,000, with manufacturers projecting to achieve VSAT costs as low as $1,000 in the near future. Much of the reduction in terminal costs has come through the use of large-scale, integrated circuit chips for the digital portion. Large cost items for a terminal are the uplink power amplifier (even with the use of a single chip device) and its installation.

Direct Broadcast Satellites (DBS)

Another major advancement for communications has been the development of effective compression techniques for transmitting video. MPEG-2 compression allows one 24 MHz satellite channel to distribute six TV channels instead of just one. Using high-power Ku-band satellites, digitally-compressed TV can be transmitted at a low cost directly to small, customer-premises terminals equipped with 18-inch antennas. These terminals are being produced in the millions and sold as a consumer electronic item for approximately $200. The terminals, which are receive–only and do not require a power amplifier for the uplink, achieve this price using a single chip to demodulate and decode the compressed video signals. The entire set-top box is composed of only two application specific integrated circuit (ASIC) chips plus one memory chip. The resultant direct TV service by Hughes' DirecTV and Echostar's DISH is very competitive with that provided by the cable TV industry. With over 13 million subscribers in the United States at the end of 1999, the DBS success has shown the potential for two-way VSAT terminals to reach a low cost.

Ka-band, Spot Beam Satellites – A Paradigm Shift

Ka-band satellites represent the biggest single change to the VSAT industry since it began in the mid-1980s. There are two principal reasons. First, unlike the traditional, bent-pipe satellite, Ka-band satellites will eventually be configured with onboard switching. This will eliminate the need to create private networks using a customer-owned or shared-hub terminal. The Ka-band satellite is a system that can potentially allow a consumer to buy a VSAT from a local retailer, point it at the satellite, and be on the network in a short time.

The second reason is that Ka-band geostationary satellites allow terminal sizes to be reduced to approximately 2 feet for an uplink TDMA burst rate of 384 Kbps using a 1-watt, 30 GHz power amplifier, and a downlink TDM rate of 90 Mbps. The 1-watt power amplifier can be produced using a single monolithic microwave integrated circuit (MMIC). With this size terminal, costs are further reduced. More importantly, however, the installation can be readily made without significant facility modifications or municipal government objection, which is necessary in order to address the consumer or mass market. To reduce the cost of a VSAT to below a $1,000, the number of units in an order need to be between 100,000 and one million. Orders for VSATs today are in the range of 10,000 or less, which really doesn't facilitate the use of high-volume production techniques.

It should be realized that the spot beam techniques used with ACTS could be applied at any RF frequency. However, the use of narrow spot beam systems at many frequencies is not possible because they would interfere with existing systems. For instance, it would be highly desirable to use ACTS-type technologies with Ku-band, whose frequencies impose significantly less rain attenuation than Ka-band. This is difficult to accomplish on both the up- and down-link for GEO Fixed Satellite Services with a user terminal smaller than 36 inches, without interfering with adjacent satellites in the geostationary orbit.

ACTS Terminals

An integral part of the ACTS program has been to demonstrate the wide variety of services that can be provided using Ka-band, spot beam satellites. Described in this chapter are the ACTS terminals and their operational performance. ACTS terminals were developed for fixed satellite services, mobile satellite services, propagation measurements, and network/satellite control.

Since the ACTS is only a test bed for proving the technology, the number of terminals produced was very limited. As a result, low-cost, high-volume production techniques were not used in the manufacturing of the terminals. The ACTS terminal program, however, demonstrates how on-demand, interactive, integrated services can be provided using very small aperture terminals that have the potential to be produced at low cost. Since the commercial ACTS-type satellites will support a large number (approaching a million per satellite) of terminals, they will be produced in high volumes similar to that for consumer electronics.

ACTS Fixed-Satellite Service Terminals

The three, fixed-satellite service terminals for ACTS are the 14-inch ultra small aperture terminal (USAT), the 4-foot T1-VSAT, and the 11-foot high-data-rate (HDR) terminal. Both the USAT and VSAT are small terminals that can be easily installed on the users' premises. The USAT operates in a fixed-beam, FDMA access mode through the spacecraft microwave switch matrix, with two-way communication provided between the USAT terminal and the 15-foot link evaluation terminal (hub station) located in Cleveland, Ohio. The T1-VSAT is configured to operate in the ACTS' TDMA demand-assigned multiple access (DAMA) environment using the onboard, base band processor (BBP) and hopping beams at data throughputs up to 1.792 Mbps. The HDR uses the microwave switch matrix (MSM) and hopping beams at data throughputs up to 622 Mbps.

73

ACTS T1-VSAT Terminal with 4-foot Antenna

Designed and manufactured by Harris Corporation.

T1-VSAT Terminal

The ACTS satellite uses two narrow hopping beams at a frequency of 19.4 GHz on the downlink and two beams at 29.2 GHz on the uplink to create the T1-VSAT network. Data from the satellite is burst-modulated on the downlink at a rate of 110.592 Mbps uncoded and 55.296 Mbps coded. On the uplink, the terminal burst rate is 27.648 and 13.824 Mbps for uncoded and coded bursts, respectively. Coded operation is used for a downlink reference burst sent at the beginning of each dwell to all terminals, and for those uplink and downlink traffic bursts suffering significant rain fade. The satellite beams are electronically hopped between all locations during a 1 ms TDMA frame. During each 1 ms frame, user data is collected from and delivered to each terminal, along with orderwire information that controls the terminal access to the network. The Master Control Station (MCS) handles the coordination and control of the network.

The T1-VSAT was developed by the Harris Corporation [50, 51] to function automatically without operator intervention and to provide two-way circuits in 64 Kbps channel increments up to a maximum of 28 channels (1.792 Mbps). The outdoor unit contains the up and down converters, the low-noise amplifier, the transmitter, and a 4-foot offset-fed parabolic dish. Prodelin made the antenna. For low-link margin areas, such as those provided by the spacecraft's steerable beam antenna, an 8-foot antenna is used. In many ways the terminal resembles the current generation of Ku-band VSATs. Some unique features are described here.

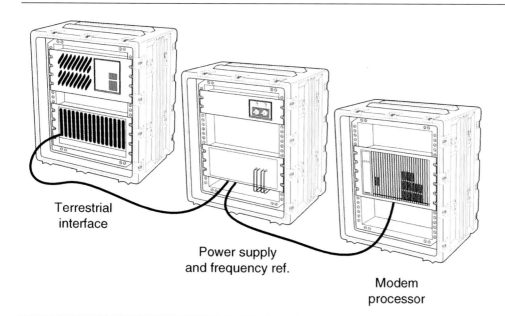

Terrestrial interface

Power supply and frequency ref.

Modem processor

ACTS – Indoor Unit for T1-VSAT

30 GHz Transmitter The most direct approach for the transmitter would have utilized a 30 GHz solid-state or traveling wave tube amplifier to provide the necessary gain and power. When the terminal was developed, Ka-band commercial activity was non-existent, so 30 GHz power amplifiers represented a significant development effort and a high cost. As a result, the solution chosen was a solid-state frequency doubler driven by a 50-watt Ku-band traveling wave tube amplifier to produce an output power of 12 watts and a total gain of 45 dB. The frequency doubler consisted of a power divider, four frequency-doubling microwave diodes, and a power combiner. Technology has progressed sufficiently enough that a totally solid-state amplifier is today's best choice. This is demonstrated by the fact that commercial Ka-band systems under development are taking the solid-state approach for 10-watt output power. To reduce the need for costly high-power amplifiers, the terminals for the current Ka-band systems transmit in a continuous low-data-rate stream, rather than the higher burst rate TDMA approach used by ACTS.

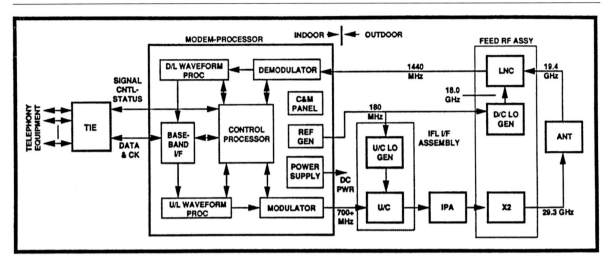

T1-VSAT Functional Block Diagram

Adaptive Rain Fade Compensation The spacecraft/VSAT link design is based on the requirement that uplink bit error rate (BER) performance shall not degrade below 5 x 10^{-7} in rain fades of up to 15 dB (the end-to-end BER requirement is 1 x 10^{-6}). To meet the fade requirement, the links were designed with a fixed 5 dB of uplink signal margin and 3 dB of downlink signal margin, augmented with adaptive rain fade compensation. The ACTS adaptive rain fade compensation is the process whereby a VSAT's data channel BER performance is automatically enhanced by burst rate reduction and error correction coding during a period of signal fade due to rain attenuation. The protocol is adaptive in that it includes a decision-making process so that fade compensation is implemented only when needed. This allows the sharing of the small pool of spacecraft-decoding capacity among terminals in all beams.

When a terminal experiences a rain fade that exceeds the compensation threshold setting, the uplink and downlink burst rates are automatically reduced in half to 13.75/55.0 Mbps. The affected burst is coded with a rate 1/2 convolutional code. This rain fade compensation provides 10 dB of additional link margin that results in average service availability in the United States of 99.5%. When the fade decreases enough to cross a cessation threshold, the coding is removed and the transmission burst rate is restored to its non-faded value. During this process, there is no interruption to the user's service.

Because the area covered by a spot beam is 120 miles in diameter, only a portion of the user terminals within that area should encounter significant rain fades. Therefore, the fade for each terminal is sensed individually.

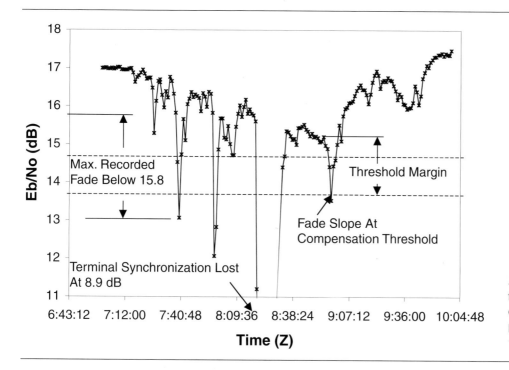

20 GHz rain fade for VSAT# 7 at Clarksburg, Maryland on 1/20/95.

Seamless Satellite/Terrestrial Network The VSAT terrestrial interface is a commercial, off-the-shelf, Redcom modular switching peripheral (MSP). The MSP is a small programmable central office, which is configured by the user with plug-in cards. The MSP supports a variety of Bell Standard hardware interfaces. Its control software has been custom-designed to provide protocol conversion between the terrestrial and the VSAT network protocols, thus enabling seamless, on-demand, integrated voice, video, and data services. Clock rate differences between terrestrial circuits and the space segment are absorbed by elastic (plesiochronous) buffers at the input and output ports to the TDMA processor. The MSP control software supports full mesh voice services with automatic number location (ANL) protocols, full and fractional Tl services, National Communications System priority services, and ISDN services.

T1-VSAT Operational Performance

Nineteen T1-VSATs, including seven units made more rugged for the U.S. Army, have been installed and operated throughout the United States, Colombia, Brazil, Equador, and Haiti. The VSATs met all requirements for providing on-demand, integrated services as described in Chapter 2, "Satellite Technology." This section centers on unique performance aspects introduced by on-board switching.

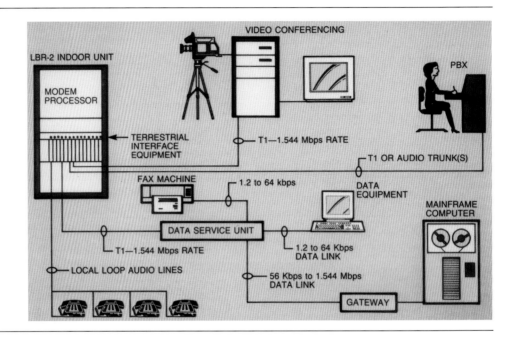

Terrestrial Compatible Communication Interfaces for T1-VSAT

Prelaunch User Terminal, Spacecraft, and Master Control System Test The ACTS BBP operation represents a new satellite mode in which the user terminals, the spacecraft, and the master control station are all required to meet a complex set of network and TDMA protocols. Prior to launch, a combined system test involving the user terminal, the spacecraft, and the master control station was performed for over three months on a two-shift per day basis to debug the software and hardware systems. This system test, which uncovered many problems prior to launch (and is not normally done with a bent pipe transponder spacecraft), proved to be invaluable. Many of the problems uncovered during the test were fairly easily fixed on the ground, but would have been much more difficult to diagnose and fix after launch. When the first post-launch BBP operation was initiated, user terminals successfully acquired the network and provided services. This was an amazing accomplishment, considering the complexity of the BBP communications system that requires nanosecond timing coordination between the hopping beams, the BBP, the user terminals, and the master control station.

User Terminal to Spacecraft Timing The timing accuracy required for a 27.5 Mbps TDMA burst to arrive at the spacecraft is +/-60 nanoseconds, or approximately +/-2 bit times. This timing accuracy is needed to enable on-board switching at each individual 64-bit word (64 Kbps circuit) boundary. The timing is accomplished by using a very accurate spacecraft clock with the BBP continually feeding back to each user terminal, whether or not their bursts are being received early or late. With the simple BBP early/late feedback, a timing accuracy that is less than this requirement is easily met. The commercial BBP systems under development have TDMA uplink burst rates of T1 (1.544 Mbps) or less. With these lower burst rates, the ACTS timing accuracy is adequate to perform operations where the normal burst preamble for bit- and unique-word synchronization can be eliminated, increasing the communications efficiency.

Adaptive Rain Fade Compensation Performance The results of the ACTS adaptive rain fade compensation are very significant for future commercial Ka-band systems. Many of the new Ka-band commercial satellites use many low-rate, uplink channels which are digitally demultiplexed, demodulated, and decoded onboard the satellite prior to switching at base band. These new satellites normally use a concatenated error correction code, consisting of a convolutional inner code and a block outer code, to achieve the desired BER with a reasonable availability. Decoding of the convolutional inner code, however, increases the processing speed and power in direct proportion to the code rate. For a rate 1/2 code, this is a factor of two. Therefore, adaptively including the convolutional code on only those few transmissions that are encountering significant rain fade can save significant spacecraft power. This is the approach taken with ACTS.

Analysis of many rain fade events for ACTS T1-VSATs has been performed [43, 44], and shows that the adaptive rain compensation system is very reliable. In no case did the system fail to successfully implement the fade compensation once the fade implementation threshold was crossed. With ACTS, it takes about one second to implement the compensation from the time the threshold is penetrated. During the one second implementation period, even the most severe rain increases the fade by less than 0.1 dB. Once the rain fade compensation is incorporated, the total link margins available are nominally 15 on the 30 GHz uplink and 13 dB on the 20 GHz downlink. For most fades, the end-to-end BER is maintained at less than the requirement of 1×10^{-6}. In the United States, the 15/13 dB link margins are adequate to maintain the BER at or below 1×10^{-6} for 99.5 % of the time.

High Data Rate (HDR) Terminal

Combining its hopping beams, wideband transponders, and microwave switch matrix, the ACTS satellite has the capability to support a gigabit satellite network (GSN). HDR terminal development by BBN Systems and Technologies (which has since been purchased by GTE) of Cambridge, Massachusetts, and Motorola Government and Space Technologies of Chandler, Arizona, was jointly sponsored by NASA's Lewis Research Center (LeRC) and the Computing Systems Technology Office of the Defense Advanced Research Projects Agency (DARPA) to create such a network [46]. The GSN system requirements include SONET compatibility, user throughput up to 622 Mbps, limited terminal transportability, and full mesh network connectivity using satellite-switched, time division multiple access (SS-TDMA). The satellite network, which, from the end users' point of view, has been designed to replicate the functions of terrestrial SONET-based fiber networks, was described in Chapter 2, "Satellite Technology."

The HDR terminal was designed to fit into a trailer (approximately 20'L x 8'W x 10.5'H). The bulk of the trailer is used for storage of the 11-foot antenna and waveguide. A small air-conditioned compartment in the front of the trailer houses the electronics.

The indoor electronics (inside the trailer) consist of the digital terminal, the burst modem, the up and down converters, and the preamplifier and power supply for the 100-watt, 30 GHz TWTA. The actual TWT is mounted outdoors on the antenna boom. Waveguide, power, and control cables are run through a trailer bulkhead, into a raceway, and out to the antenna. The decision to put the majority of the electronics inside, and to use waveguide to and from the antenna, was a compromise between providing climate-control for the electronics and dealing with problems introduced by the long waveguide run.

The digital electronics used to control each HDR terminal support SS-TDMA transmissions, and provide over-the-satellite, cross-connect, SONET-switching capability. TDMA bursts are constructed as a sequence of satellite cells preceded by a short preamble. Each satellite cell is composed of 648 bytes of data and a 48 byte Reed-Solomon checksum. The satellite cells contain a fixed number of bytes and are either 8 or 16 μs in duration, depending on whether the burst rate is 696 or 348 Mbps. The length of each burst is in multiples of the 32 μs slot period. The difference between the burst rate of 696 (or 348) and the user rate of 622 (or 311) is due to TDMA overhead and the incorporation of Reed-Solomon 232/216 block coding to achieve a 1×10^{-11} BER. Bursts arrive at ACTS phase locked to the 1000 slot space-time-switching plan (32 μs frame). Tight synchronization of all HDR terminals with the satellite's

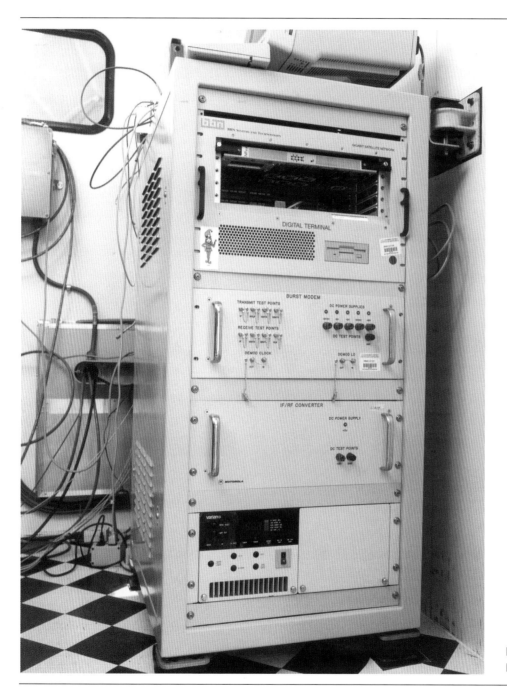

HDR Indoor
Equipment Rack

onboard switching and beam-hopping plans allows very short guard time
between bursts and simple satellite tracking.

ACTS High-Data-Rate (HDR) Ground Station

Manufactured by Motorola and BBN.

Development Risks The principal technical risks for HDR terminal development were associated with the 696 Mbps burst rate. This rate presented significant challenges for the burst modems and the digital processor. The architecture and design approach for the digital processor is provided in reference [52].

The key modem technology developed for the ground station included fast acquisition circuitry for the AGC and the carrier/clock recovery [46]. The final preamble length achieved for the TDMA bursts was 1 μs.

The extremely hard-limiting and finite (albeit large) bandwidth of the ACTS transponder presented difficulties in the implementation of the modem. The hard limiting was incorporated into the transponder channel to equalize the levels of signals in the ACTS microwave switch matrix (MSM). The intent of this was to reduce the effect of leakage between channels due to the finite isolation of the switch FETs used in the MSM. The bandwidth of the

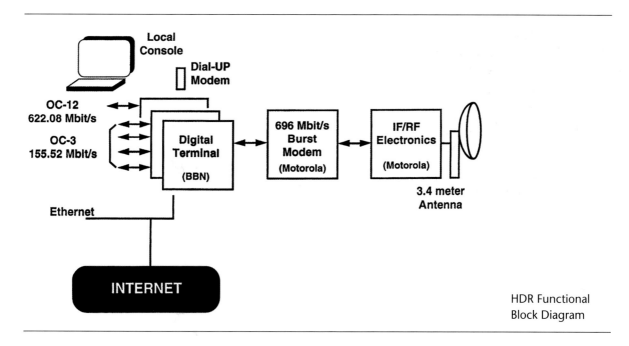

HDR Functional
Block Diagram

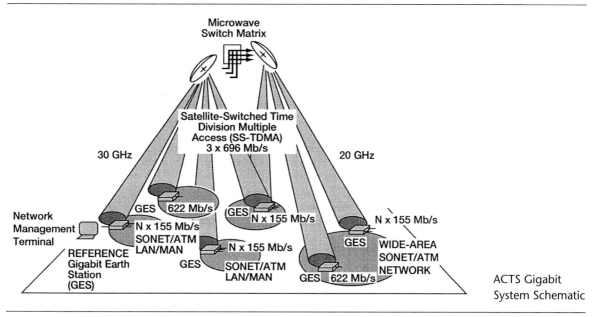

ACTS Gigabit
System Schematic

spacecraft channels is reasonably flat over about 800 MHz, and rolls off rapidly beyond that. Inter-symbol interference distortion due to hard limiting can be minimized by the use of constant-envelope modulation. This was demonstrated in the use of 220 Mbps modulation in the link evaluation terminal [53]. For the HDR modem, staggered BPSK and QPSK modulation was used and filters were made as wide in bandwidth as possible to provide a close-to-constant envelope.

A further source of inter-symbol interference were the long (36 feet) waveguide runs to and from the antenna, which are responsible for group delay variations of almost 3 nsec across the downlink band (19.2–20.0 GHz), and approximately 2 nsec across the uplink band (29.0–29.8 GHz). There is considerable deterioration in inter-symbol interference when using SQPSK at the 696 Mbps rate. The group delay variation is remedied with RF equalizers on both the uplink and downlink. Each equalizer consists of a cascade of three waveguide band-pass filters, designed to provide complementary group delay to that of the waveguide.

Ultra Small Aperture Terminal (USAT)

Commercial Ka-band satellites being developed with onboard, base band switching have terminal uplink rates ranging from 128 to 384 Kbps and downlink rates on the order of 30 to 90 Mbps. One of the benefits of onboard, base band switching is that the uplink and downlink rates can be asymmetric. This allows, for example, video broadcast on the downlink and low-rate-data on the uplink for single hop communications. This results in a small terminal with a 20 GHz wideband receiver and a low-power, low-cost 30 GHz solid-state transmitter. NASA's Lewis Research Center developed the USAT to demonstrate the potential capability of future commercial Ka-band satellite systems with this type of terminal.

The USAT consists of the following elements:

- a 14- or 24-inch offset-fed antenna (manufactured by Prodelin)

- 30 GHz solid-state power amplifiers ranging from 1/4 to 4 watts (the 1/4 and 1.0 watt were developed by LNR)

- 4.0 dB noise-figure receivers by Miteq and Electrodyne

- a 70 MHz user interface

- the necessary up and down converter equipment for frequency translation to Ka-band [54]

Ten were assembled using commercially available components. The two stage upconverter chain and 2 watt, 30 GHz amplifier, which were made from MMIC chips, fitted into a 5" x 3" x 1" box mounted on the feed arm. In the future, a large number of this terminal type will be produced for commercial systems, using application specific integrated circuits (ASIC) and monolithic microwave integrated circuits (MMIC) to reduce cost.

Ultra Small Aperture Terminal (USAT) – 14-inch Antenna, with Richard Reinhart of NASA Lewis Research Center

ACTS Mobile Satellite Service Terminals

Three different types of mobile terminals have been developed for use with ACTS. Each type is classified according to its antenna—a rotating reflector, a mechanically-steered slotted waveguide, and a phased array. The terminals were used in numerous land vehicle, ship, and aircraft operations described in Chapter 4, "User Trials." Even with the ACTS' narrow spot beams with gains in the neighborhood of 50 dBi, the mobile terminal antennas must have significant EIRP to provide satisfactory low-data-rate (2.6 Kbps) communication. As a result, personal communication using a handheld terminal with an omni antenna is not feasible for ACTS and was not attempted. All these terminals operate in a fixed-beam, FDMA access mode through the microwave switch matrix with two-way communication provided between the mobile unit and a 4- to 18-foot hub station.

85

ACTS Mobile
Rotating Reflector
Antenna

Rotating Reflector Terminal

This mobile terminal was primarily used in conjunction with a land vehicle to communicate through the satellite with a fixed hub terminal and was developed by the Jet Propulsion Laboratory (JPL) [55]. Using frequency division multiple access (FDMA), an unmodulated pilot signal is transmitted from the hub station to the mobile terminal. The mobile terminal uses the pilot to aid in antenna tracking and as a frequency reference for Doppler offset correction and pre-compensation. For system efficiency, the pilot signal is only transmitted in the forward direction. Therefore, the setup for this terminal includes two signals in the forward direction: the pilot signal and the data signal. In the return direction (mobile terminal-to-ACTS-to-fixed station), only the data signal is transmitted. Operational data rates for this mobile terminal are nominally 2.4, 4.8, 9.6, 64, and 128 Kbps.

The antenna is the critical Ka-band technology item developed as part of this terminal design. A passive, elliptical reflector-type antenna is used in conjunction with a separate high-power amplifier (TWTA with 10 watts output RF power) to produce a fan beam in elevation. The antenna provides a minimum EIRP (on boresight) of 23 dBW, a G/T (also on boresight) of -5 dB/°K, and a bandwidth of 300 MHz. The maximum EIRP possible is 30 dBW. The reflector

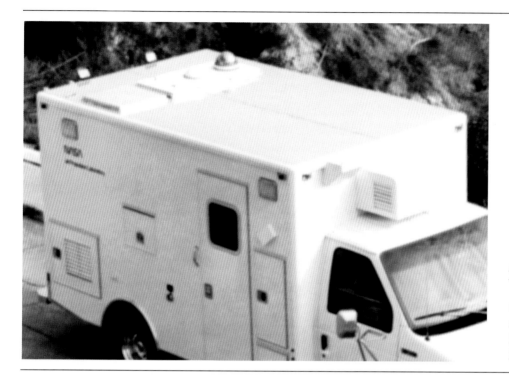

ACTS Mobile
Terminal

*Designed and
manufactured by Jet
Propulsion
Laboratory (JPL).*

resides inside an ellipsoidal, water-repelling radome with an outside diameter of approximately 8 inches (at the base) and a maximum height of approximately 4 inches.

The azimuthal pointing system enables the antenna to track the satellite for all practical vehicle maneuvers. This antenna is mated to a simple, yet robust, mechanical steering system. Pointing was accomplished by dithering the antenna in azimuth about its bore site at a rate of two cycles per second. The pilot signal strength measured through this dithering process is used to complement the inertial information derived from a simple, turn-rate sensor. The combination of these two processes was designed to keep the antenna aimed at the satellite even if the satellite was shadowed, for up to ten seconds. This mechanical pointing scheme is one of the benefits of migration to the Ka-band frequencies. The considerably smaller mass and higher gain that is achievable relative to L-band make the mechanical dithering scheme feasible and eliminate the need for additional RF components to support electronic pointing. The necessary processing resides in the antenna controller.

The baseline modem implements a simple, yet robust differentially-coherent BPSK (DPSK) modulation scheme, with rate 1/2, constraint length 7 convolutional coding and interleaving. This choice of modulation scheme was dictated largely by concerns over the performance impact of phase noise onboard the satellite. The performance specification of the modem, including implementation losses, is a bit error rate of 1 x 10^{-2} at an E_b/N_0 of 7.0 dB in an additive white gaussian noise (AWGN) environment. Essential to the modem design is the built-in capability to withstand deep, short-term shadowing. The modem free wheels (i.e., does not lose synchronization), through signal outage caused by roadside trees, and will reacquire the data after such a drop out. The modem has also been designed to handle possible frequency offsets due to Doppler and other frequency uncertainties of around 10 kHz changing at a maximum rate of 350 Hz per second.

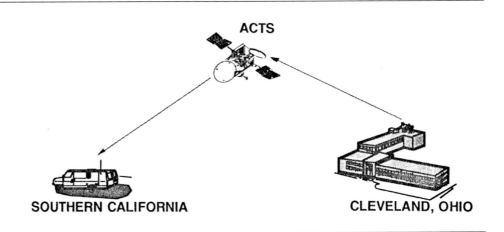

Baseline AMT System Performance Test Configuration

SOUTHERN CALIFORNIA **CLEVELAND, OHIO**

Rotating Reflector Terminal Performance The mobile terminal was tested in a fixed position using ACTS to determine how well its demodulator and decoder performed. A bit error rate (BER) of 10^{-3} was achieved at an E_b/N_0 of 6.8, 6.7, and 8.5 dB, for data rates of 9.6, 4.8, and 2.4 Kbps, respectively. It was known well in advance of any tests that significant signal fading or blockage would occur due to obstructions from buildings, utility poles, and trees. Moving vehicle tests verified the capability of the antenna system to successfully free wheel between deep fade or blockage periods experienced in a suburban area. When an obstacle shadows the satellite line-of-site for a period greater than 30 seconds, such as when driving behind a building, the antenna loses alignment with the satellite. When the obstacle is cleared from the satellite view, the antenna automatically re-initiates signal acquisition.

Although the JPL terminal was primarily used for demonstrating communication applications (see Chapter 4, "User Trials"), propagation data was also taken using the ACTS mobile terminal [56]. Johns Hopkins University and the University of Texas also prepared a mobile ground vehicle solely for the purpose of collecting propagation measurements at 20 GHz, using the ACTS steerable beam antenna. Extensive propagation data was collected in Maryland,

Texas, and Alaska [57], and is reported in Chapter 5, "Ka-band Propagation Effects."

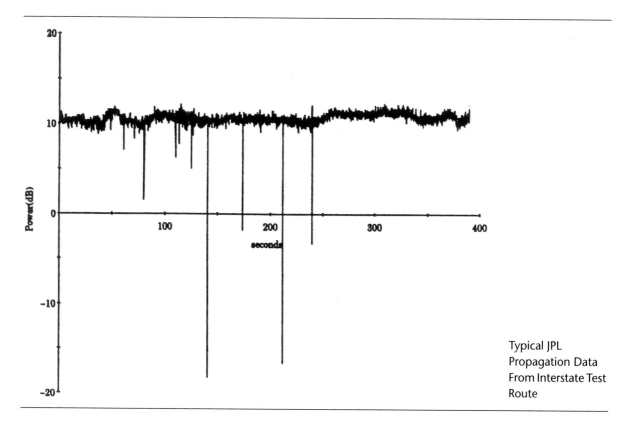

Typical JPL
Propagation Data
From Interstate Test
Route

Mechanically-Steered Slotted Wave Guide Terminal

The ACTS broadband mobile terminal was developed by NASA's Jet Propulsion Laboratory (JPL), together with various industry/government partners, to investigate high-data-rate mobile applications of ACTS technologies [55]. The terminal was used on a C-141 aircraft, a Saberliner 50 aircraft, an Army HMMWV vehicle, an oil exploration ship, and the U.S.S. Princeton. These operations demonstrated the viability of Ka-band aeronautical and mobile satellite communication at bi-directional Tl data rates. Present day commercial aeronautical satellite communication systems are only capable of achieving data rates of tens of kilobits per second. The use of the Ka-band for wideband aeronautical communication has the advantage of spectrum availability and smaller antennas, while avoiding large rain attenuation by flying above the clouds the majority of the time.

The heart of the broadband mobile terminal is the mechanically-steered, slotted waveguide antenna. This high-gain, mobile antenna employs an elevation-over-azimuth pointing system to allow it to track the satellite while the

89

aircraft, vehicle, or ship is maneuvering. EMS Technologies, Inc., developed the antenna and radome. The antenna design utilizes two slotted waveguide arrays, and is mechanically-steered in both azimuth and elevation. The polarizer in front of each array achieves the required circular polarization.

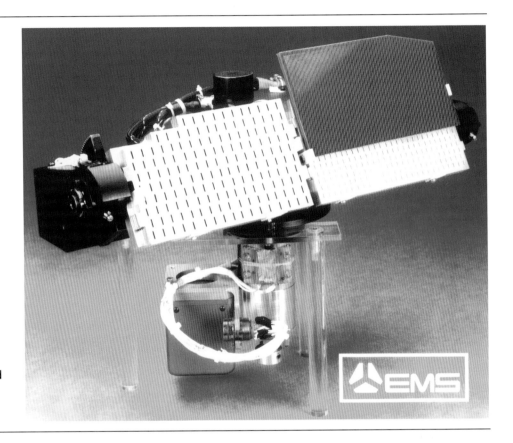

Slotted Waveguide
Antenna with
Polarizers Removed

*Manufactured by
EMS Technologies.*

The antenna radome was designed for low RF loss, with the mechanical integrity to withstand the aerodynamic loads of a jumbo jet. The radome is shaped with a peak height of 6.7" and a diameter of 27.6"—roughly the size of the SkyRadio radome currently flying on United Airlines and Delta Airlines aircraft. Antenna installation requires a 3.5" diameter protrusion into the fuselage to allow the necessary signals to pass to and from the antenna.

The antenna is capable of tracking a full 360 degrees in azimuth and –50 degrees to zenith in elevation. The antenna has a transmit gain of 30 dBi and a receive-sensitivity (G/T) of 0 dBi/°K which are greater than those for the rotating reflector terminal described earlier. The actual dimensions of the transmit and receive array apertures are each approximately 4" x 8", and the arrays are approximately 0.5" thick. The polarizers add another 0.3" to the total antenna thickness.

ZWM.8ESS

Title: The Advanced Communications
Technology Satellite: An
Insider's Account of the
Emergence of Intera...
Cond: Good
Date: 2024-10-11 23.15.36 (UTC)
mSKU: ZWM.8ESS
vSKU: ZWV.1891121111.G
unit_id: 18810253
Source: CATALINA

...m was required to maintain pointing ...ut all phases of flight. The antenna posi- ...muth mechanism with a precision of a ...itioner is controlled by a tracking algo- ...formation: a three-axis inertial rate sen- ...n input; and pilot signal strength feed- ...atellite spot beam. The IRS, with 50 Hz ...antenna assembly, provides the major- ...cking system. The overall tracking sys- ...to 60°/sec and 30°/sec/sec in azimuth, ...1. The transmit side of the array is pow-

ZVW.1891121111.G

delist unit# 18810253

xxxxx

Antenna Radome
on NASA's C-141
Aircraft

Antenna Radome on
Rockwell's
Saberliner 50

Mechanically-Steered Slot Waveguide Terminal Performance Rockwell Collins outfitted a Saberliner aircraft with the terminal, and performed various tests to determine how well the antenna system tracked the spacecraft for normal maneuvers. From taxiing on the ground prior to take-off to the ascent (including banking and ending at cruise altitude), the aircraft's received pilot signal-to-noise ratio peak-to-peak variation was less than 1.5 dB. This performance was judged to be quite good.

Active Phased-Array Terminal

Active phased-array antenna systems hold great promise for meeting the demands of future satellites with a beam(s) that can be steered rapidly to any location. For mobile or fixed-LEO user terminals, they offer the additional potential advantage of being a slim, flat plate shape or contoured to fit the surface of a vehicle or aircraft. To date, most success in active arrays has occurred at frequencies lower than the Ka-band. A cutting-edge technology effort was undertaken by NASA's Lewis Research Center (LeRC) and the Air Force to develop 30 and 20 GHz GaAs MMIC devices that could form an active phased-array antenna.

Each transmit element consisted of a separate 30 GHz MMIC phase shifter, followed by an amplifier. Each receive element consisted of a separate 20 GHz MMIC low-noise amplifier, followed by a phase shifter. A single 30 GHz transmit array was developed by LeRC and Texas Instruments. Three 20 GHz receive arrays were developed in a cooperative effort between the Air Force's Rome Laboratory and NASA, and took advantage of existing Air Force array development contracts with Boeing and Lockheed Martin (previously General

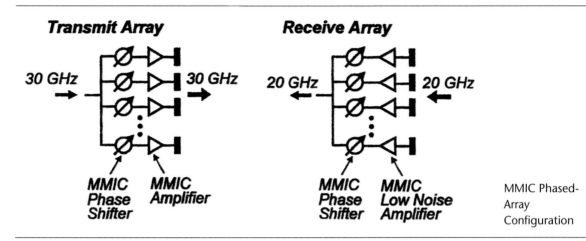

Transmit Array

30 GHz → → 30 GHz

MMIC Phase Shifter   MMIC Amplifier

Receive Array

20 GHz ← ← 20 GHz

MMIC Phase Shifter   MMIC Low Noise Amplifier

MMIC Phased-Array Configuration

Electric). These active phased-array antenna systems were mounted on ground vehicles and aircraft to demonstrate their capability to provide voice, video, and data links through ACTS to a fixed hub terminal. Because the array antennas were limited in extent and experimental, these developments were considered an initial step toward the eventual development of practical systems for Ka-band application.

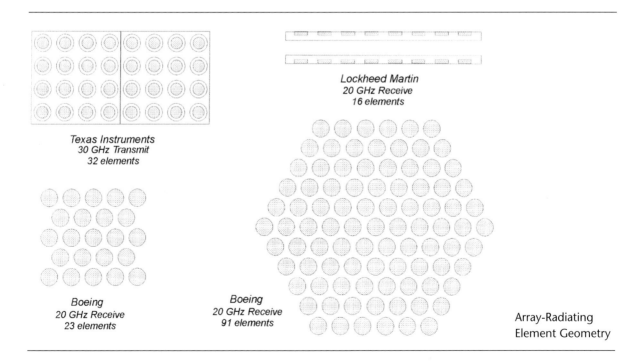

Lockheed Martin 20 GHz Receive 16 elements

Texas Instruments 30 GHz Transmit 32 elements

Boeing 20 GHz Receive 23 elements

Boeing 20 GHz Receive 91 elements

Array-Radiating Element Geometry

Texas Instruments 30 GHz Transmit Array The Texas Instruments (TI) 30 GHz transmit array has a total of 32 elements and consists of two, 16-element 4 x 4 modules, mounted in a protective experimental housing [58, 59]. The TI approach features a thin, tile architecture in which the components are mounted in a plane perpendicular to the antenna boresight. Each of the two 16-element sub-array modules is 3.2 cm x 3.2 cm x 0.75 cm thick. The array element spacing of 0.8 of a wavelength supports scanning up to ± 30° without grating lobes. The module design is based on a hybrid integration approach in which conventional wire bonding is used for interconnecting MMIC devices to the signal distribution layers.

Sixteen RF lines each feed a 4-bit MMIC pin diode phase shifter of switched-line length type and a three-stage, 100 mW PHEMT power amplifier and is electromagnetically coupled to a cavity-backed, aperture-coupled, circular patch element. Logic commands for selection of individual phase bits and amplifier drain bias control (on/off) are transmitted via a serial data bus to a custom application specific integrated circuit (ASIC), which converts the serial data to parallel data for simultaneous control of the MMIC devices. A thermoelectric cooler and a small fan in the antenna housing provide thermal stability.

The measured output power for each MMIC power amplifier is approximately 100 mW, with 20 dB of gain. For the subarray module, the roll-off in gain—as a function of scan angle—was measured to be less than 3 dB over a 30° scan angle. At boresight, the measured EIRP for the array is 23.4 dBW.

Texas Instruments 30 GHz Transmit Array

20 GHz Receive Arrays Although the Lockheed Martin and Boeing designs involve fundamentally different packaging concepts, both use a brick architecture in which the active components are mounted in modules parallel to the boresight of the radiating elements. The radiating elements have a nominal 1/2 wavelength separation supporting scanning to +/–60° without grating lobes.

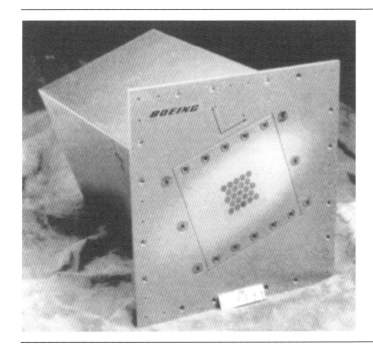

Boeing 20 GHz
Receive Array
– 23 Elements

Array Mount for USAF C-130
JWID-95 Exercises

Boeing 20 GHz Receive Array The small 20 GHz receive array built by Boeing is a brick architecture, but uses a different design concept than Lockheed. The array consists of a cluster of 23 active channels on a triangular grid with a half wavelength of separation. Each channel is a separate, dielectrically loaded, circular waveguide into which a MMIC LNA and 4-bit phase shifter are inserted. The LNAs initially have a gain of 18 dB and a 9 dB noise figure. In addition to the active components, each module has DC circuitry and a logic chip for phase shifter control. In this array, the RF outputs of the 23 channels are combined in a multi-mode space feed. No active cooling is provided. The measured G/T of the array is -16.6 dB/°K at boresight.

The large Boeing 20 GHz receive array, which consists of 91 elements, is essentially a larger (more elements) version of the 23-element array and it has significantly improved RF performance. The LNA devices used in this larger array have a noise figure of 2.5 dB. As a result, the measured G/T is –4.5 dB/°K at boresight and –9.2 at a 70° scan angle.

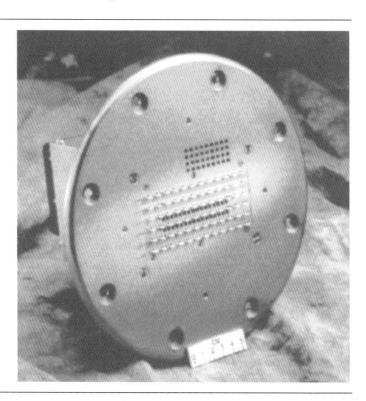

Lockheed Martin 20 GHz Receive Array – 16 Elements

Lockheed Martin 20 GHz Receive Array The small 20 GHz receive array built by Lockheed Martin consisted of 16 dipole antenna elements arranged on a 2 x 8 rectangular grid with half wavelength separation. Two plug-in cards (or trays), each having 8 active receive channels, form the array. Each active channel consists of a printed circuit dipole antenna, connected by a microstrip to a

PHEMT low-noise amplifier and a GaAs 3-bit phase shifter. Each channel has a logic chip for phase-shifter control. The overall gain of each channel is 23 dB with a noise figure of 3 dB. The RF output of the 8 channels is combined in an 8:1 beam-former and amplified by a follower amplifier. The RF output of each tray is combined in a 2:1 beam-former. A fan cooled the array. The overall gain of the array from the array face to the final beam-former was measured to be 42 dBi. The measured G/T of the array is -16.1 dB/°K at boresight.

Texas Instruments 30 GHz transmit array and Boeing 30 GHz receive array mounted in the roof of an Army's HMMWV vehicle.

Phased-Array Controller Open-loop steering was chosen after careful consideration of the antenna beam widths in combination with gyro accuracy, and by restricting the rate of attitude changes to a maximum of 10° per second. A single software control loop processes position updates from the global positioning satellite (GPS) once per second under interrupt control. It also processes current gyro attitude (roll, pitch, yaw) measurements via an analog-to-digital converter board. Combining the GPS position information, current attitude information, and the known position of the ACTS satellite, the proper steering angles are calculated and the antennas electronically steered in the desired direction. Antenna updates occurred from 0 to 18 times per second under flight conditions that maintained acceptable link performance. While initially designed for the aeronautica terminal demonstration, the controller met the moving ground terminal (HMMWV) requirements as well.

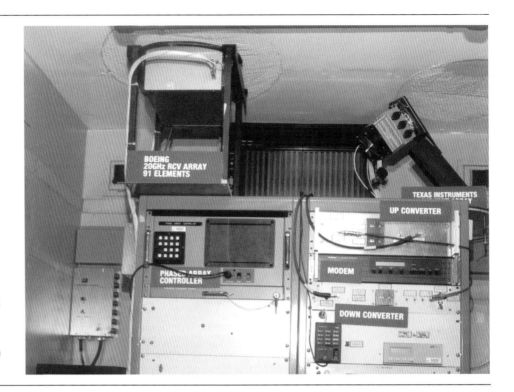

Phased-array inside HMMWV vehicle. RF windows are circular elements in top of enclosure.

Both aeronautical and ground vehicles tests were run using the phased arrays to determine how well the antenna systems performed for both commercial and military applications. Even with the simple array controller, the antenna transmit and receive beams successfully tracked the ACTS satellite and provided useful, two-way, low-data-rate communication. The successful performance of experimental, proof-of-concept MMIC Ka-band arrays in field demonstrations indicates that high-density MMIC integration at 20 and 30 GHz is indeed feasible. The results of this program created a strong incentive for continuing the focused development of MMIC array technology for satellite communications applications, with an emphasis on packaging and cost issues.

Network and Spacecraft Control

Two major components of the ACTS ground system are the NASA ground station (NGS) and the master control station (MCS), collocated at NASA's Lewis Research Center. Both were developed by COMSAT Laboratories [60, 61]. The NGS provides the RF communication links with which the MCS performs its various network control and monitoring functions for the base band processor mode of operation. In addition the NGS RF communication links are used by Lockheed Martin's spacecraft control center, initially located in East Windsor, New Jersey, to receive telemetry information from the spacecraft and to transmit commands to the spacecraft. In 1998, this spacecraft control center was moved to Newton, Pennsylvania. The NGS/MCS combination also serves as a traffic terminal for communication data being transmitted in the base band processor mode.

NGS communication links with the spacecraft are conducted at Ka-band frequencies. When the link margins are exceeded by a rain fade, the communication network and the command and control capability for the spacecraft ceases. At that time, the spacecraft operates autonomously.

The NGS 18-foot antenna system built by TIW provides nominal EIRP of 74, 68, and 77.5 dBW for the BBP 110 Mbps, the BBP 27.5 Mbps, and the S/C command uplinks, respectively. Similarly, the NGS G/T for the BBP 110 Mbps and telemetry downlinks is 30.3 and 30.7 dB/°K, respectively.

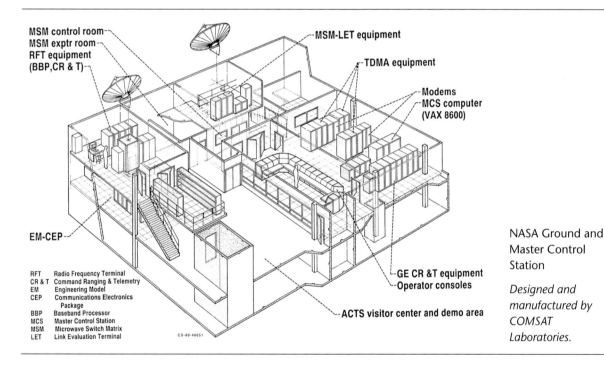

NASA Ground and Master Control Station

Designed and manufactured by COMSAT Laboratories.

For command, ranging, and telemetry (CR&T) functions, the dynamic range for telemetry reception is 14 dB (100 kHz bandwidth). The command uplink uses two data rates—5 Kbps or 100 Bps. At the normal operating EIRP of 79 dBW, the command link marginsare 18 and 27 dB, respectively.

Using the BBP 110 Mbps up- and downlinks, the MCS/NGS communicates with the spacecraft to perform its BBP network control functions. These include sending BBP programming instructions to the spacecraft, receiving BBP status information, and sending and receiving orderwire messages for the user terminals. For these functions, the link performance is specified at a BER of 1E-06 end-to-end, which is partitioned into 5E-07 each for the up- and downlinks. The approximate clear-air link margins (above 5E-07) are 14 and 20 dB for the up- and downlinks, respectively. Rain fade compensation (rate 1/2 FEC coding plus 50% reduction in transmission rate) adds 10 dB to each of these margins.

In summary, when the uplink rain fade exceeds approximately 28 dB, the BBP network will crash and the CR&T transmission will also be impaired. For the three years from 1995 through 1997, the number of BBP network rain outages each year were 17, 24, and 4, respectively. The duration of each network outage was primarily a function of how fast the operator manually brought the system back up.

Propagation effects impact the NGS antenna pointing. This phenomenon is well known for Ku-band step-tracking systems, and its potential for Ka-band is greater. In actual NGS operations, propagation anomalies can disrupt antenna-pointing, requiring operator intervention to restore operations. This effect is not chronic, and tracking is generally reliable during light fades. During heavy rain fade events, however, the antenna occasionally mispoints. The effect is common enough that antenna tracking is manually disabled when severe weather is expected.

18-foot NGS antenna (right) and 15-foot link evaluation antenna (left) located at the Lewis Research Center in Cleveland, Ohio.

On-Demand Circuit Capacity

The MCS is responsible for the real-time control and monitoring of the ACTS BBP communication networks, as well as the associated control of the spacecraft payload—including the BBP. The ACTS architecture utilizes the features of the BBP to provide a highly flexible, high-performance network, providing the user with circuit capacity on demand in any increment of 64 Kbps. In addition, the architecture automatically provides rain fade compensation for any terminal in the network.

To perform its dynamic network control functions, the MCS accepts capacity requests from user terminals via the inbound orderwire (IBOW), then formulates and sends the appropriate assignment messages to user terminals via the outbound orderwire (OBOW). It also formulates BBP control messages and sends them to the BBP via the BBP control orderwire. The transmission of these messages is time-coordinated to accomplish synchronized changes of the input memories, routing switch, output memories of the BBP, and traffic bursts arriving from, and received by, the user terminals. Consequently, demand assignment changes occur without any interruption to the services being carried.

COMSAT Development Team: (left to right) Dave Meadows, Michael Barrett, William Schmidt – program manager, Sam Kouvaris, Lou Parker, William Fallon, Steve Struharik, M. Miller.

The DAMA functions performed by the MCS employ several algorithms, depending upon the circuit required (single/multichannel or single/multidestination). Development of these algorithms required consideration of conflicting objectives and constraints, including response time, frame utilization, BBP operational constraints, recovery from errors in control messages, experimental flexibility, and implementation cost. In particular, the requirement to provide call setup times on the order of 3-5 seconds and the complexities of programming the BBP offered significant technical challenges. Several simulation programs were developed and utilized to test and refine alternative approaches to the DAMA problem.

The MCS programs the BBP using microcode-level instructions transmitted to the BBP via the 576 Kbps BBP orderwire channel. Because of the need to reprogram the BBP approximately every three seconds, a feed-forward protocol is used on the orderwire channel, eliminating the need for time consuming command acknowledgment. This protocol is designed to ensure that the BBP is reliably programmed, even in the event of bit errors in the command channel. Characterization of the orderwire channel performance during system tests showed that the orderwire circuits (both the traffic terminals and the BBP) were viable even with rain fades of up to 25 dB. At this point, the traffic circuits experience a BER > 1E-02, and thus are unusable, but the control circuits permit continued stable network operation.

System timing is an important consideration. The MCS contains a cesium-beam reference oscillator, which provides the fundamental timing reference for the BBP network. The long-term stability of this oscillator is approximately 1 part in 2.5 x 10^{12}, consistent with Bell System standards. The MCS continually compares BBP onboard clock-drift to this frequency standard and adjusts the BBP clock accordingly.

COMSAT Operations Team: (front row, left to right) Terry Bell, Thomas Tanger, Steven Struharik, Lloyd Blackman, and Douglas McGlamery. (back row, left to right) Charles Sheehe, Bruce Curry, Jeffrey Glass, Timothy Mazon, and Kamara Brown.

For running and controlling the BBP TDMA network, the MCS employs a low-performance VAX 8600 with a total of 200,000 lines of executable C-language code. The on-demand network control, although very complicated and sophisticated, was implemented using a fairly modestly sized program. In the network control center, one to two people are normally required to operate the BBP network.

Spacecraft telemetry data acquisition and processing equipment located at the Lewis Research Center in Cleveland, Ohio, with Richard Krawczyk of NASA.

Lockheed Martin Operations Team: (left to right) Steve Cohen, Don Cooley, Gene Callahan, Kent Mitchell, Dave Jarvi, Al Phebus, and Darnell Moye. Val Barron, the ACTS operations manager, was absent when the picture was taken.

The role of the ground operations team in the success of the ACTS program cannot be overemphasized. The team's knowledge of the spacecraft and ground equipment allows for the compensation of a number of shortcomings that tend to creep into even the most carefully conceived, tested, and executed system.

Propagation Terminals

ACTS provides beacons at 20.2 and 27.5 GHz for use in making attenuation measurements. The NASA ACTS propagation program was designed to obtain slant-path attenuation statistics for locations within the United States and Canada for use in the design of Ka-band communications satellites [62]. Experimenters at seven different locations (British Columbia, Colorado, Alaska, Maryland, New Mexico, Oklahoma, and Florida) collected propagation data for more than five years. Critical to these measurements was the design and calibration of the propagation terminals [63, 64]. All measurement sites had identical hardware and software, and used the same calibration procedure to produce consistent data with an absolute attenuation measurement error of less than 1.0 dB. This high fidelity database is available on CD-ROM and is being added to the ITU worldwide attenuation database.

The beacon signals at each frequency are collected using a 1.2 meter offset-fed Prodelin antenna, sent to a low-noise amplifier, down-converted, split, and sent to a beacon digital receiver and a total power radiometer. All RF components, digital receivers, and radiometers are enclosed in temperature-controlled housings to maintain the desired accuracy and stability.

The digital receiver, which converts the beacon analog signal using a 12-bit converter, was developed to overcome the difficulties previous experiments had with analog systems. The shortcomings of analog systems included lack of stability, long acquisition times, and difficulty in manufacturing and maintenance. The beacon's signal power is normally reported at 1 sample per second, but for scintillation studies the data is reported at 20 samples per second. The radiometer receives its signal after many stages of amplification. As a result, the radiometer must be calibrated every 15 minutes by switching-in a reference load.

After the raw data is collected, the beacon signal level and are processed to apply the calibration information and remove the beacon-level offsets, as well as an estimate of atmospheric absorption. The output of the calibration process is the beacon attenuation referenced to free space, sky noise temperature, the radiometer attenuation, the weather sensor, and other system status measurements. The propagation terminals produce attenuation data with an accuracy of better than 0.3 dB over a dynamic range greater than 20 dB. It should be noted that the attenuation measurements include the effects of antenna-wetting, which

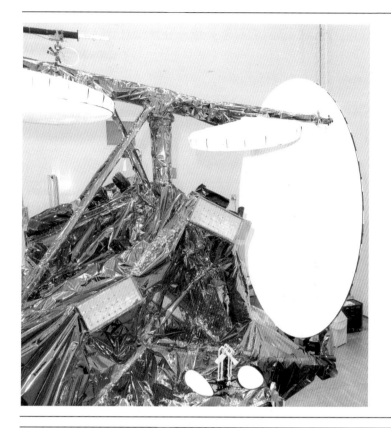

20 and 30 GHz beacon antennas located in the bottom of the spacecraft picture.

Propagation terminal for measuring rain attenuation at 20 and 30 GHz. Seven identical terminals are located throughout the United States.

becomes significant at Ka-band frequencies. The seven terminals have been very reliable with the terminal downtime being less than 1% of the total measurement period.

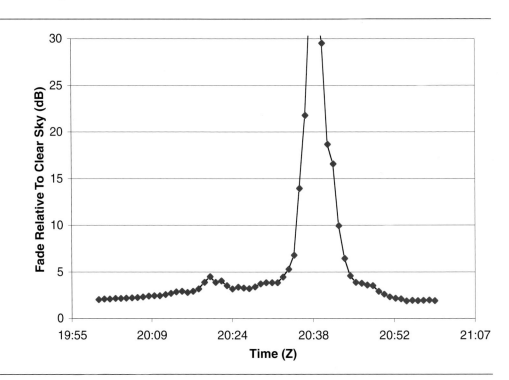

20 GHz beacon attenuation at Maryland site (1/15/95).

Measured by COMSAT Laboratories.

Concluding Remarks

The ACTS terminals described in this chapter helped prove the utility of digital communications at Ka-band for fixed, land mobile, shipboard, and aeronautical applications from a single satellite.

The terminals have collected useful propagation data, demonstrated the feasibility of dynamic rain fade compensation, promoted the use of phased-array and other small, lightweight antennas, achieved high-data-rate, fiber optic quality transmissions, and shown that ultra small aperture terminals (USATs) are feasible for customer premises service. Finally, the ACTS USATs have been implemented with highly integrated MMIC components similar to what is required for mass production and the achievement of price levels acceptable to consumers.

CHAPTER 4

USER TRIALS

From the very beginning, ACTS was intended to be an in-orbit test bed for validating both advanced Ka-band technologies and on-demand, integrated voice, video, and data services that would be needed for the twenty-first century. As such, the user program was an important element, and U.S. industry played a key role in working with NASA to formulate it. NASA recognized that the full potential of ACTS technology could only be realized if private industry assumed an active role in the conduct of technology validation and application trials. The development and flight validation of both the advanced technologies and new, cost-effective services has allowed private industry to adapt these technologies and services into their individual commercial systems at minimal risk. As it turned out, ACTS development was very timely. The large market increase for integrated services requires that satellite systems provide digital communication on an on-demand basis that are fully compatible with terrestrial networks.

Background

The user program had its beginnings back in 1979, when the ACTS program was formulated [65]. NASA, in planning its future satellite communication program, formed a carrier working group (CWG) from various industry organizations that provided commercial terrestrial and satellite services (see Chapter 1, "Program Formulation"). The CWG was charged with helping NASA formulate the technology, flight system, and user-trial requirements, and providing overall guidance and advice. One of the key activities of the CWG was to identify various types of experiments and user trials needed to characterize and validate the new spacecraft and ground technology being planned. In June of 1980, NASA issued the Experiments Planning document. This document contained a listing of 73 potential experiments submitted by members of the CWG and NASA, as well as a summary of each, and was useful in finalizing the requirements for ACTS. In March 1983, NASA issued a formal Notice of Intent (NOI) for Experiments. The NOI solicited expressions of intent to conduct experiments or user trials. Eighty-five organizations responded to the NOI, representing a broad cross-section of the U.S. telecommunication industry, government agencies, and universities. From these responses, 122 experiments and user trials were proposed. The response was used to obtain support for an ACTS flight program. As an example, Mr. Don Nowakoski testified before Congress that Western Union would invest $5 M in ACTS earth terminals to conduct user trials.

After the ACTS contract was awarded in 1984, a broad outreach and solicitation effort was initiated to make potential users aware of the capabilities of the ACTS system and offer an opportunity to become involved. Several working focus groups were established. Regional meetings were held throughout the country in an effort to reach key industry, government, and university organizations. The purpose of these activities was to solicit involvement and foster cooperative user investigations and activities.

Industry, government, and university experimenters meet at the Lewis Research Center.

Between 1985 and 1993, yearly ACTS experiment conferences were held to provide a detailed overview of ACTS spacecraft and ground terminal development and update experimenters and others on the progress and status of the experiments program.

In December of 1988, NASA formed an Industry Advisory Group. The function of this group was to advise NASA on the various elements of the ACTS program:

• the experiments program plan

• the process and criteria for selecting experiments

• ways to foster the telecommunication community's involvement

• approaches to providing experimental earth stations, and

• types of experiments and demonstrations to be performed

Members of the advisory group included Al MacRae from AT&T; Luis Diaz from Citibank; Len Golding from Hughes Network Systems; Otto Hoernig and

Jay Ramasastry from CONTEL/ASC; Jack Keigler from RCA; Bob Drummond and Al Bartko from the Defense Communication Agency; Troy Ellington from GTE; Joe Campanella from COMSAT; John Geist from Harris; Rob Briskman from Geostar; Marvin Freeling from GE American Communications; William Garner from American Mobile Satellite Corporation; Bert Edelson from Johns Hopkins University; John Clark from the U.S. Naval Academy; William Pritchard from Pritchard and Company; and John McGill and Charles Russell from ALASCOM. Dean Olmstead, the NASA program manager for ACTS, was instrumental in fostering user trials that were strongly supported by the industry and proved the commercial viability of the ACTS communication technology.

In August of 1991, NASA issued an Experiment Opportunity Announcement (EOA). The EOA solicited proposals for conducting investigations that would test and evaluate the key technologies of the ACTS flight and ground systems, as well as various application services. Forty of these were initially selected to be ACTS investigations. By December of 1999, a total of 104 investigations involving over 120 different industry, government, and university organizations had been accepted. In addition to the investigations, over 100 different demonstrations had been conducted.

Categories

The primary goals of the program were to:

- conduct a complete set of technology verification experiments that validated and characterized ACTS technology

- conduct a balanced set of user investigations and application demonstrations that evaluated on-demand, integrated voice, video, and data applications, and

- collect a comprehensive series of propagation measurements to aid in the design of future Ka-band communication satellite systems

This chapter covers the user investigations. Chapter 2, "Satellite Technology," presents significant results from the technology verification experiments, and Chapter 5, "Ka-band Propagation Effects," provides detailed information on the propagation measurements.

User applications comprised over half of the investigations conducted. They involved a variety of fixed, mobile, and video broadcast services. Most of the user trials were oriented toward services with commercial potential and included the following: medical; terrestrial network restoration; business, science and ISDN networks; education; DOD tactical communications; broadcast

Members of the user development team from LeRC, NASA headquarters, and Lockheed Martin.

Left to right: Ron Schertler, users manager; Laura Randall; Dean Olmstead, NASA program manager from December 1988–March 1992; and Richard Gedney, LeRC project manager.

video; supervisory control and data acquisition (SCADA); very high-data-rate SONET/ATM networks; aeronautical, land vehicle, and maritime mobile; and protocol and network interoperability.

Operations

Operations were initiated on December 6, 1993, after the completion of all spacecraft, ground system, and network in-orbit checkouts. The ACTS system offered considerable flexibility in accommodating multiple simultaneous and independent users in either the BBP, MSM, or mixed modes of operation. Reconfiguration of the spacecraft payload between the various modes of operations was accomplished in less than 30 minutes. During the fall and spring equinox periods, the spacecraft solar panels were eclipsed up to 72 minutes per day. During this time, the payload was shut down. Except for these eclipse shutdowns, experiment operations were conducted 24 hours per day, seven days a week, 365 days per year. Geostationary operations continued until July of 1998.

Inclined Orbit Operations

The ACTS spacecraft was designed for four years of operation in orbit, and was loaded with sufficient hydrazine stationkeeping fuel to provide that length of service. With a continued strong demand for experiment time and interest by industry in continuing to use ACTS as a demonstration vehicle, NASA commissioned Lockheed Martin to investigate the feasibility of operating in an inclined orbit long before the expected depletion of the hydrazine. Studies confirmed the feasibility, and NASA directed Lockheed to implement the necessary autonomous control software to provide accurate pointing with the increased inclination. Meanwhile, NASA equipped the ground stations with tracking capabilities to follow the daily excursions of the spacecraft above and below the equator. The modifications were carried out in the proper sequence to ensure the continuity of the experiments.

After four years and nine months, full stationkeeping was discontinued in July 1998. With full stationkeeping, ACTS remained positioned at 100° west longitude and was maintained within a +/-0.05° box. With no north-south stationkeeping, orbital inclination increases at a rate of 0.76° per year. Using only a very small amount of hydrazine for east-west stationkeeping, ACTS currently remains geosynchronous at 100° west longitude +/-0.05°, with orbital eccentricity near zero. In this inclined orbit mode, ACTS operations are planned to be extended until June 2000, when the remaining hydrazine fuel will be used to raise the orbit of the spacecraft above the geostationary altitude and all onboard systems and subsystems will be turned off.

Applications Investigations

As a forerunner to future commercial communication satellites, the ACTS test bed offered an extremely versatile platform on which to conduct a wide range of user service applications. In addition, ACTS demonstrated the capability to interoperate with the terrestrial networks providing users with worldwide connectivity. This fits the telecommunication needs of today's global economy that requires interconnectivity among all countries. ACTS has operated with a network of over 70 terminals of various types for both fixed and mobile services throughout North and South America and Hawaii. Data rates using these various types of user terminals ranged from 2.4 Kbps up to 622 Mbps. Many of the terminals were integrated with fiber optic networks to form a hybrid satellite/terrestrial network to demonstrate, validate, and accelerate the role of satellites as key components on the information superhighway.

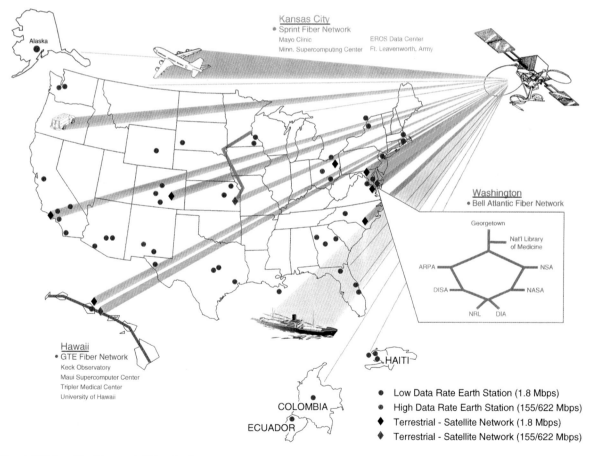

The ACTS Satellite/Terrestrial Test Bed

For instance, the high-data-rate terminals were integrated with the Bell Atlantic fiber network in Washington, D.C.; the Sprint fiber network in Kansas City, Kansas; and the GTE fiber network in Hawaii. Many of the T1-VSAT earth stations, as well as the various types of USAT and mobile terminals, were also connected into the public, switched terrestrial networks.

Some of the user services that were investigated in the user program are not yet available or effectively provided on a broad scale by today's commercial communication satellites. ACTS offered the potential to help spur the introduction of these new services and provided new ways to reach a larger number of users at a lower cost than possible with pre-ACTS technology. While the applications experiments and trials were being run, engineering teams were carrying out technical validation experiments, evaluations, and characterizations on the ACTS communication links and networks.

The following sections provide a comprehensive overview of many, but not all, of the various user tests and trials using the ACTS system. They should showcase the tremendous versatility of the ACTS system and the wide range of users and organizations that participated in the ACTS user program.

Medical

Fourteen and one-half percent (14.5%) of our GNP is spent on health care. In spite of this, many of our citizens (especially in rural areas) do not have easy access to quality medical care. Providing cost-effective health care to patients in isolated or remote areas far from medical centers is both a national and a global concern. ACTS provided high-quality, wide bandwidth communication links to demonstrate telemedicine (i.e., remote medical diagnostics) and consultation. The capability of ACTS to use communication circuits on-demand has the potential of making telemedicine cost-effective.

Mayo Clinic

Led by Dr. Bijoy Khandheria, physicians from the Mayo Clinic in Rochester, Minnesota conducted extensive telemedicine trials over ACTS [66, 67] between 1994 and 1996. In one phase, they used T1-VSATs to perform remote medical diagnosis and evaluation of patients in the Pine Ridge Indian Health Services Hospital on the Lakota reservation in Pine Ridge, South Dakota. The 45-bed Pine Ridge hospital provides health services to the reservation. The hospital had 15 positions for full-time primary care physicians and 87 nurses. Historically, it had been difficult to fill the 15 full-time positions, and physician turnover tended to be high. Because of the remote location of the reservation, consulting and continuing education services were difficult to obtain for the health care professionals at Pine Ridge. There are close to 2,500 hospitals with limited medical personnel like Pine Ridge Clinic that serve about 24% of the U.S. population. T1-VSAT service provided Mayo physicians in Rochester, Minnesota, with real-time voice, video, and data connections with patients on the reservation hundreds of miles away. Physicians, including specialists, were able to see and talk to the patients as well as receive information instantly from such instruments as electronic stethoscopes, ultrasound scanners, and electrocardiogram recorders.

In one case, a Mayo physician evaluated a child with skin lesions (who had been treated for years without a cure) and determined the diagnosis to be leprosy. Treatment was prescribed via ACTS and the patient was cured! During another session, a Mayo psychiatrist examined a patient via ACTS who was paranoid, prone to seizures, alcoholic, an inhalant abuser, and who showed Parkinson-like symptoms. The Pine Ridge medical staff considered this one of their most challenging cases. Through the process of remote examination, the consulting psychiatrist was able to provide keen insight and a correct diagnosis. The Mayo team carried out over 50 different clinical consultations with the Pine Ridge Indian Reservation, involving 13 different medical specialties. In addition, the team conducted physician education and training programs for health professionals. Most participants—doctors, nurses and patients—reported that the quality of the T1 video and audio signals was good and acceptable for delivery of basic educational and health services.

In other trials, the Mayo team in Rochester was connected by terrestrial fiber to the ACTS HDR earth station in Kansas City and via ACTS to another HDR earth station at the Mayo Clinic in Scottsdale, Arizona. The Mayo Clinic was interested in demonstrating the clinical and technical feasibility of using combined satellite, terrestrial, and local hospital networks to provide on-demand, short-duration access between remote hospitals and tertiary care centers. The medical areas addressed were angiography, echocardiology, family medicine, and radiology. Again, numerous consultations and diagnoses were conducted, this time using X-ray, magnetic resonance imaging (MRI), ultrasound, and computed tomographies (CT), in addition to patient-doctor sessions and expert-group consultations. The average file size of the transmitted images was 300 megabytes. The data rate used over ACTS was 155 Mbps, which allowed an image to be sent to the remote specialist in less than one minute. The Mayo network consisted of a satellite and terrestrial links, and the hospital's local area networks at both facilities. The teleradiology experiment alone involved 13 Sun workstations connected into the hospital network for collecting, transmitting, receiving, and reviewing the MRI and CT images.

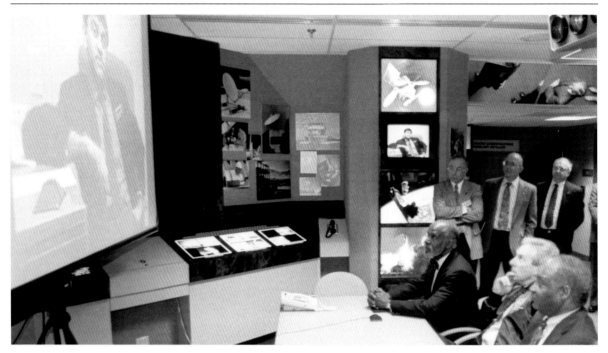

Dr. Bijoy Khandheria of the Mayo Clinic briefs NASA administrator Dan Goldin over the ACTS link on Mayo's telemedicine experiments. Don Campbell, director of the Lewis Research Center, and Dr. Wesley Harris, NASA's associate director of aeronautics, are seated with Mr. Goldin.

Conducting over 100 different trials, Mayo demonstrated the capability of ACTS to deliver high-quality imagery in real time over a hybrid network system. Mayo's overall conclusion was that the use of ACTS-like technology for real-time, interactive, quality healthcare delivery was feasible, with a high degree of acceptance on the part of both doctors and patients. As a result of the ACTS trials, the Mayo's catheterization and echocardiography laboratories have started programs for doing remote interpretations.

Krug Life Science

In another telemedicine trial, Krug Life Science and NASA's Johnson Space Center (JSC) in Houston, Texas, examined patients at the Fitzsimmons Army Medical Center in Boulder, Colorado, using the T1-VSAT links. Using a portable, high-resolution, retinal-imaging camera, Houston specialists examined live color images of the retinas of 53 patients in Boulder, in realtime.

Remote diagnoses made by the consulting specialists in Houston were in complete agreement with those made by the Army ocular specialists in Boulder. As a result of these successful demonstrations at T1 rates, JSC developed a

T1 network of medical consultants to provide support during shuttle operations, as well as during astronaut training in Russia.

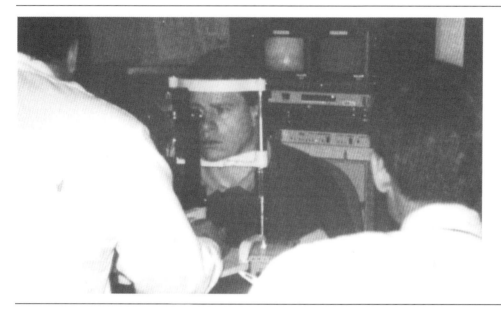

Medical personnel conduct a retinal examination on a patient in Boulder, CO. Video imagery from the ophthalmic instruments was transmitted to NASA JSC via ACTS for remote diagnosis.

Billings, Montana, was the site of another telemedicine trial. In this test, an ACTS USAT was coupled with a telemedicine instrument package (TIP)—developed by Krug Life Science and NASA JSC to monitor the health of astronauts in space. The TIP is a suitcase-size, portable diagnostic center that includes an electrocardiogram, blood gas content detector, electronic stethoscope, and a special camera for dermatology examinations. Although designed and developed to be used on the shuttle and the space station, the TIP can be utilized on earth—particularly in rural medicine facilities or remote workplaces such as off shore oil platforms—to improve the quality of, and access to, medical care.

Using the TIP and a LeRC-supplied, 23-inch diameter USAT operating at T1 (1.544 Mbps) rates, doctors at St. Vincent Hospital in Billings conducted various staged trauma trials and demonstrations with patients at the Exxon Corporation refinery and the Crow/Northern Cheyenne Hospital. The Exxon refinery, located in the Billings area, has an occupational medical clinic. The Crow/Northern Cheyenne Hospital is a 24-bed rural hospital located in south central Montana.

Doctors were pleased with the diagnosis quality, high-resolution video, audio, and data transmission provided by ACTS and the TIP. The results demonstrated the ability of rural health care professionals to use satellite technology to send a patient's vital information to a hospital hundreds or thousands

Krug Life's
Telemedicine
Instrument
Package (TIP)

of miles away for proper evaluation by qualified medical staff. In many cases, such telemedical evaluations not only provided expert consultation, but also alleviated the need to transport a patient to a regional hospital—saving unnecessary travel as well as undue stress to the patient.

The Cleveland Clinic and the University of Virginia

The Cleveland Clinic, the University of Virginia, and NASA's LeRC teamed up to investigate the use of satellite-based telemammography [68]. Breast cancer is the second leading cause of cancer-related deaths among American women, although it is 90% curable if detected early enough. Breast cancer screening through mammography is recommended for tens of millions of American women. In 1996, an estimated 56 million U.S. women were of the recommended age for annual mammography screening. Mammography requires skilled and experienced radiologists—who are usually located in large medical facilities—to interpret the images. People in rural, low-density population areas generally have no direct access to such expertise.

T1 (1.544 Mbps) satellite links can provide affordable connectivity for those patients, allowing direct and immediate access to mammography experts. Using satellites, the possibility exists to greatly improve mammography screening, especially in remote areas where patients might have to travel several hours for their annual screening. Whereas current mammography films are generally shipped to an interpretation center (resulting in days or weeks before the results are known), the potential for near real-time mammography screening via satellite means that a patient can receive results right after the screening. In the 7 to 10% of the cases requiring follow-up, patients are immediately available—thus eliminating the requirement for a revisit to the screening center.

Originally, the ACTS T1 links alone could not provide adequate capacity for fast transmission of the large files associated with mammography, so the team decided to investigate advanced data compression techniques. Large patient files of up to 320 megabytes resulted from multiple views, high image resolution requirements, and the need to compare current to previous breast images. At a T1 rate, it would take 27 minutes to transfer all the files. However, by using data compression techniques, the transmission time can be drastically reduced while still allowing for quality diagnosis. So far, more than 5,000 digitized mammography images have been transmitted over ACTS—all have been received without error. Most of these images were mammography film images scanned at 100-micron resolution. Some were scanned at 50-micron resolution.

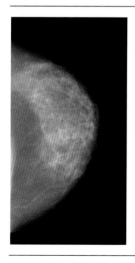

Mammogragh image sent via ACTS. This is a typical mammographic digital image that was originally scanned at a resolution of 100 microns. The resolution has been reduced here to make it easier to print.

These transmitted images were compressed (using wavelet image compression) at ratios of 8:1 to 30:1. Some uncompressed images were also transmitted. Radiologists at the Cleveland Clinic have been able to detect the effects of using image compression down to ratios of 16:1 when viewing the decompressed images on a high-resolution gray scale monitor. One radiologist could detect it at 12:1.

The Cleveland Clinic has performed clinical receiver operating characteristic (ROC) studies of the 8:1 compressed, satellite-transmitted images. Using a blind study of 60 cases (where each case consisted of four images), the satellite-transmitted, 8:1 compressed images were compared to the original film images. The result was that diagnostic accuracy of the compressed, transmitted images was found to be equivalent to the original film. With 8:1 image compression, a set of four mammographs can be transmitted in about 3 minutes, which is acceptable for near real-time analysis. Additional tests are continuing with patients at the Ashtabula County Medical Center in Ashtabula, Ohio, whose images are transmitted to specialists at the Cleveland Clinic.

By making use of the T1 rate connections, the basic image transmissions and remote diagnosis can be augmented by teleconference capability to provide teleconsultation with remote physicians, or directly between patient and specialist. Satellite telemammography alone represents a potential market of tens of millions of telemedicine sessions for future commercial satellite service providers.

EMSAT and JPL

EMSAT and JPL successfully demonstrated the use of the ACTS mobile terminal (AMT) for emergency medical communications [69]. The experiment highlighted the feasibility of mobile satellite communications to provide better pre-hospital communication than was available with current terrestrial radio technology. It evaluated the transmission and reception of satellite digital voice for two-way, pre-hospital communication, one-way transmission of patient data from field paramedics to the base hospital, and telemetry of patient assessment data to the base hospital. These trials simulated communications with paramedics at the accident scene and en route to the hospital. Voice and data transfers were tested at 2.4, 4.8, and 9.6 Kbps rates. Results indicated that satellite communication for Emergency Medical Service (EMS) units is possible and desirable. For each data rate, communications were clear and usable—a measurable improvement over current radio frequency communications.

University of Hawaii, Georgetown Medical Center, and Ohio Supercomputer Center

The University of Hawaii (UH), the Georgetown University Medical Center (GUMC), and the Ohio Supercomputer Center (OSC) teamed in a DARPA-sponsored trial to test, evaluate, and demonstrate the use of the ACTS gigabit network in radiation-dose treatment planning [70]. This trial was really an illustration of how communication links can provide for a collaborative effort among remote centers of excellence.

Radiation therapy has been the primary treatment for cancer for decades. The idea behind radiation therapy is to deliver a dose of radiation to the tumor while minimizing doses to the surrounding normal tissue. The results of the treatment depend greatly on the planning. To kill deep-seated tumors, the radiation must be cross-fired from a number of different angles, all aligned to intersect the tumor. Treatment planning assures that the beams do intersect the tumor. Conventional treatment planning is performed on two-dimensional slices of CT images and is generally not adequate. Three-dimensional (3D) treatment planning, which considers the three-dimensional structure of the patient's anatomy, has recently emerged as a better method. Such planning, which allows 3D beams, delivers a more conformal dose on the tumor and offers the promise of a more effective treatment. However, 3D planning is currently very restrictive due to the limited access to powerful supercomputers. The computer requirement involves 3D computations and 3D graphics rendering. It also requires a high-performance supercomputer to allow physicians to view anatomy, draw beams, and evaluate dose distributions. A high-speed computer workstation is then required to compute and even optimize dose distributions based on the beam characteristics and anatomy information.

ACTS gigabit satellite networks allow remote supercomputer centers to perform the necessary computations and 3D graphics visualizations in near realtime, which can then be used for interactive planning sessions with treatment physicians. In this field trial, GUMC obtained a set of MRI and CT scans from a cancer patient and sent them via ACTS at 155 Mbps to UH, where a 3D profile was developed. The 3D profile was sent to GUMC, where they developed an initial radiation beam treatment plan, which was returned to UH. Using this radiation beam plan from GUMC and the patient's MRI and CT scans, a dose computation software program was run on a UH supercomputer. From UH, the results of the dose computation were sent via ACTS to OSC where it was fused with the patient's MRI and CT data that was transmitted from GUMC. At OSC, a visualization of both the patient's imagery and the dose treatment plan was generated on a high-performance Silicon Graphics workstation. The visualization—in video form—was sent to GUMC for evaluation. Radiologists at GUMC, however, sometimes required additional iterations that were subsequently performed in near realtime at UH and the OSC.

The ACTS high-speed data links were used to connect the UH and OSC centers of excellence and their supercomputer capabilities together to perform a service for GUMC. This successful trial of distributed treatment planning via the ACTS gigabit network is expected to stimulate new medical services, as

121

well as motivate networks that transcend time, distance, and resource barriers. With high-speed networks, medical personnel will be able to share resources interactively on a global scale in real or near realtime.

Military Applications

The battlefield is dynamic and mobile, as evidenced in the Gulf War. Providing the battlefield commander with updated intelligence, targeting, weather, and command information is vital. Operations require tactical commanders to operate independently. To date, communication capabilities that provide intelligence and command orders to such tactical units have been limited to simple voice and low rate, narrow-bandwidth communications.

Army Ruggedized
T1-VSAT–
Outdoor and
Indoor Units

The Army Space Command spearheaded a number of Army organizations in the evaluation of ACTS to provide battlefield commanders with on-

122

demand, wide-bandwidth, integrated services. Seven T1-VSATs were made more durable and designed to be highly transportable for tactical operations.

The primary objectives of the Army's program were to test and evaluate ACTS in multiple field exercises to realistically simulate battlefield communication requirements. In a number of operational field trials and demonstrations with tactical units, ACTS was used to provide on-demand, integrated voice, video, data between forward tactical units, and geographically dispersed rear echelon commands. The T1-VSATs provided local commanders with the ability to dictate the assignment of bandwidth for specific user services. Some of the many Army communications needs tested were:

- video teleconferencing for command and control

- reconnaissance imagery

- integrated weather charts/imagery

- mobile phone base station relay

- telemedicine

- logistical supply

- video conferencing for morale boosting

- interconnection into the Defense Commercial Telecommunications Network (DCTN) to provide a seamless satellite/terrestrial network

Haiti Operations

The Army used its seven T1-VSATs in support of Operation Uphold Democracy in Haiti [71]. Three VSATs were located in Haiti, while others were located at Ft. Bragg, North Carolina; Ft. Drum, New York; and Army Space Command Headquarters in Colorado Springs, Colorado. At Ft. Bragg, the VSAT was tied into the Defense Communication Telephone Network (DCTN). This connection allowed units in Haiti to videoconference with military bases around the world. The VSAT also connected into the Army's own tactical radio network.

At Ft. Drum, a connection was made from the VSAT into the U.S. public, switched telephone network. This capability expanded the communications capability in Haiti to full interconnectivity with the U.S.

ACTS provided secure video for daily conferences between the Joint Task Force in Haiti and commanders back in the United States. U.S. Army Colonel James Campbell reported that battlefield commanders were able to solve unique military problems using this capability. The participation of many people at each end of the link, and the ability to see their body language,

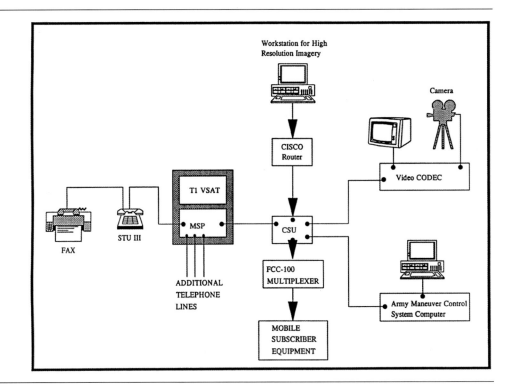

Workstation for High
Resolution Imagery

Camera

CISCO
Router

Video CODEC

T1 VSAT

MSP

STU III

CSU

FAX

ADDITIONAL
TELEPHONE
LINES

FCC-100
MULTIPLEXER

Army Maneuver Control
System Computer

MOBILE
SUBSCRIBER
EQUIPMENT

U.S. Army ACTS
Applications
Equipment
Configuration

added substantial value over voice and low-bandwidth videoconferences. In addition to supporting command operations, the ACTS videoconferencing capability was used to support morale conferences between soldiers in Haiti and their families back home in the United States. The Walter Reed Medical Center in Washington, D.C., connected via landline to the COMSAT VSAT in Clarksburg, Maryland, also provided telemedicine support to Haiti field operations. President Clinton's address to the commanders and troops in Haiti on October 6, 1994, was transmitted over ACTS when a last minute glitch interrupted the planned communication links.

Mobile Vehicle Communications

In mobile communication tests, the Army used both mechanically rotating antennas and solid-state, phased-array terminals.

Rotating Reflector Terminal In one trial, the ACTS mobile terminal (AMT) communication equipment and tracking reflector antenna shown in Chapter 3, "Terminal Equipment," were used to test and evaluate video transmissions and receptions from a moving high mobility multipurpose wheeled vehicle (HMMWV) [72]. Battlefield video imagery and control data were transmitted

124

Army T1-VSAT set up to support a training exercise at Ft. Irwin in California.

to and from the vehicle via ACTS. The smaller equipment size (especially of the antenna), and the high-data-rate capability provided by the ACTS Ka-band spot beams enabled this application. In another trial, the same equipment was mounted and tested in an unmanned ground vehicle that is currently under development by Lockheed Martin for the U.S. Army. Satellite systems have the potential to increase the range, as well as the data rate, of such unmanned, robotic vehicles.

MMIC Arrays The Department of Defense (DOD) was also very interested in the potential of the MMIC arrays to provide integrated voice, video, and data communication. The advantages of arrays (small, conformal, and with elec-

Captain Ken Ilse, U.S. Army Space Command, explains the importance of bandwidth on demand for army tactical units to NASA Administrator Dan Goldin and to Don Campbell, LeRC Director (seated lower left).

tronic steering) were showcased to the military services. The MMIC phased-array antennas, which are described in Chapter 3, "Terminal Equipment," were mounted in a HMMWV as well as a C-130 aircraft, and participated in a number of Army and Air Force exercises in a one-year period between September 1994 and September 1995 [73].

Voice, data, imagery, and slow-scan video links were established and tested via ACTS. Four different MMIC arrays [Texas Instruments, Boeing (2), and Lockheed Martin] were used in the course of these demonstrations. Each array is electronically steered. With the arrays mounted in the HMMWV, engineers from the Lewis Research Center demonstrated duplex voice communications at 9.6 Kbps using a variety of commercial and standard military communication equipment. In conjunction with the Prairie Warrior exercise at Ft. Leavenworth, Kansas, duplex voice and video transmissions were achieved to and from the HMMWV at a data rate of 21.6 Kbps using a commercially available videoconferencing system. Although the video was slow-scan at these rates, it nevertheless demonstrated the potential of mobile, interactive, video communications. In a demonstration to the U.S. Navy at Newport, Rhode Island, the MMIC arrays were used to receive a 128 Kbps video and data stream in a simulated operational situation of high-bandwidth information sent via a satellite to a submarine.

The most ambitious MMIC array demonstrations took place as part of the Joint Warrior Interoperability Demonstration in 1995 (JWID-95). Between September 26-29, 1995, at Camp Pendleton, California, links were successfully established between ACTS, a HMMWV, and a C-130 aircraft using the MMIC

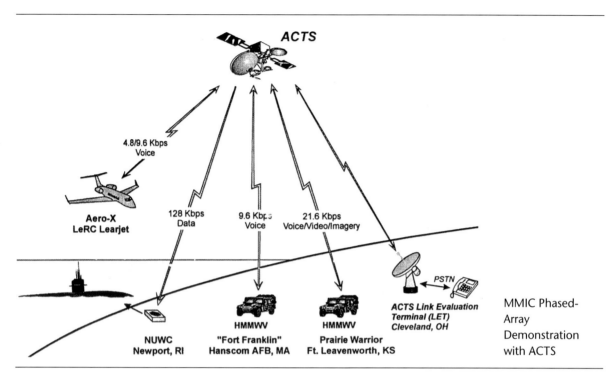

ACTS

4.8/9.6 Kbps
Voice

Aero-X
LeRC Learjet

128 Kbps
Data

9.6 Kbps
Voice

21.6 Kbps
Voice/Video/Imagery

PSTN

*ACTS Link Evaluation
Terminal (LET)
Cleveland, OH*

NUWC
Newport, RI

"Fort Franklin"
Hanscom AFB, MA

HMMWV

Prairie Warrior
Ft. Leavenworth, KS

HMMWV

MMIC Phased-
Array
Demonstration
with ACTS

arrays. In one test, 115 Kbps of encrypted image data were transmitted from Ft. Meade, Maryland, to LeRC via a Ku-band Galaxy satellite and relayed to both the HMMWV and the C-130 via ACTS. In another test, a video was transmitted at a rate of 1 Mbps from the Naval Research Lab via a second Ku-band satellite, and relayed to both the HMMWV and the C-130 via ACTS. In a final test, 16 Kbps of voice was transmitted from the HMMWV back to the Naval Research Lab.

In another demonstration of satellite-terrestrial interoperability using the MMIC arrays mounted in a HMMWV, NASA Administrator Dan Goldin, placed a call to his secretary in Washington, D.C., via ACTS and the public switched telephone network.

Active MMIC phased-arrays are of great interest to both the military and the private aircraft industry because they can provide electronic beam-steering in a compact configuration conformal to a vehicle or aircraft surface—unlike mechanically steered reflectors, which require radomes protruding above the surface. For the same reasons, proposed FSS LEO satellite systems, such as Teledesic and SkyBridge, hope to use arrays in the user terminal. The successful performance of the experimental, proof-of-concept arrays has helped create a strong incentive for continuing the focused development of MMIC array technology for satellite communication.

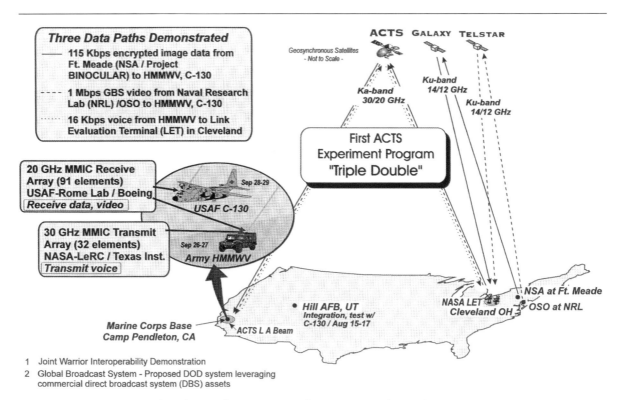

Overview of ACTS MMIC phased-array demonstrations that were part of JWID-95.

Dan Goldin talks to his secretary over ACTS using the MMIC phased-array antennas mounted in a HMMWV, while Gus Martzaklas of LeRC looks on. Phased-array antennas are mounted in the roof.

128

Business Networks

Over the past decade, practically all aspects of modern commerce have become heavily reliant on communication technology. Business networks using VSATs are among the fastest-growing applications of communication satellites. VSAT networks in the early 1990s were arranged in a star configuration, with all traffic routed through a central hub earth station. This arrangement necessitated a double hop to the satellite for VSAT-to-VSAT traffic. For voice applications, a double hop through a geostationary satellite is unacceptable because of the undesirable, long transmission delay. Using ACTS technology—with high-gain antennas and the capability to perform onboard switching—double hop is eliminated, and single hop, VSAT-to-VSAT is accommodated.[1]

National Communications System (NCS)

From the federal response perspective, there are functional requirements, outlined and managed by the National Communications Systems (NCS), to restore services for emergency operations in case of natural or man-made disasters. One of the main functions of the NCS is to provide restoration of low-data-rate (T1) communication during these disasters, and the NCS sees satellite communication as the chief means of aiding in this cause.

To this end, NCS, NASA, Mitre Corporation, and JPL planned and executed a series of simulated communication restorations using ACTS' T1-VSATs [74]. In performing the tests, terrestrial T1 connectivity between Reston, Virginia, and Pasadena, California, was initially established and then disrupted. Using special software, connectivity was manually reestablished via two T1-VSATs. For operational restoration systems, software would be developed to automatically reestablish the severed connection. The ease of implementation and the effectiveness of such a system were the key parameters evaluated. Both full T1 trunk and individual 64 Kbps circuits were tested. Another noteworthy aspect of these tests included call prioritization and preemption, where higher-priority callers usurped communication channels from lower-priority users. These tests proved highly successful. NCS concluded that the ACTS system provided high-quality, consistent, secure, and clear voice communication. Further, it was possible to maintain communication with a very low bit error rate. Establishing on-demand calls that met the NCS's security

---

1. Single hop communications between VSATs is now feasible for high-power, bent-pipe, Ku-band satellites. The advantage of onboard processing satellites is that they have the potential to provide services at a lower price.

requirements was easy and relatively efficient. Call prioritization and the use of personal identification numbers (PIN) also worked well.

Huntington Bank

The increasing dependence on communication by businesses brings with it more stringent reliability requirements for the networks. In some areas of business, such as the financial sector, government regulations require each company to prepare and test business restoration plans. Most businesses combine two strategies to provide the protection that they need: redundancy and backup systems. Frequently, satellites provide the backup.

Ohio University led an experiment sponsored by Huntington Bank that was designed to determine if satellite circuits were technically compatible with terrestrial transmission equipment and terrestrial network management systems [75]. These are important considerations if satellite networks are to be considered for redundant or backup operations. In the past, terrestrial and satellite networks have been completely separate, with each having its own set of equipment and no interoperability between them.

In one series of tests, customer transactions, ATM transactions, account balances, transfer-of-fund data, and computer-scanned check data were all transmitted over ACTS in tests of instant, on-demand switchovers. Fifteen separate bank data circuits were successfully and seamlessly routed from Sun Recovery Services in Philadelphia via a hybrid ACTS/terrestrial network. Full compatibility was found between the ACTS system and the terrestrial network. The T1-VSATs could be integrated into the Huntington Bank's T1 network in a straightforward manner without the need to develop special configurations for the terrestrial equipment. Circuit setup times on ACTS were within the range needed for redundant circuits; the times were comparable to or better than on-demand terrestrial T1. No problems were found in the cutover to the ACTS circuits. Terrestrial T1 carrier networks use a low-speed, out-of-band channel to monitor T1 circuit equipment. ACTS currently does not support this channel. All in-band network management control, however (such as that typically found in end-user T1 equipment), did function without any problems.

NBC

In another user trial, NBC-TV used T1-VSATs to transmit video between different fixed broadcast locations. Their experiment demonstrated that T1 (and even sub-T1) video and audio can be used to provide real-time video and audio feeds for network news applications. The quality of the ACTS links was such that the transmitted imagery and audio was maintained at sufficient fidelity to be broadcast directly in realtime on the network.

The ACTS mobile terminal (AMT) was also used in tests and demonstrations with NBC. Current communication capability for mobile newsgathering by ground vehicles is limited to cellular telephone service, when available. Once the satellite news gathering (SNG) van is on the scene and stationary, video can be transmitted. Using ACTS and the AMT, however, a full duplex compressed video link at data rates up to 768 Kbps was established while on the move. Interconnecting the link into the terrestrial network allowed the video to be delivered to NBC headquarters in New York City. NBC gained significant insight into satellite technology, which will help guide their development of new and exciting newsgathering capabilities.

Small Business

The commercial Ka-band, spot beam satellite systems currently under development are intended to serve small businesses, home offices, and regional offices using small user terminals—at a total throughput of 16 Kbps to 384 Kbps and sometimes larger. The communication technology is primarily for telephony, multimedia, video conferencing, file transfer, Email, and Web browsing. These specific on-demand services were performed in an integrated fashion in many different user trials using the ACTS T1-VSAT network, which illustrated to service providers the potential for spot beam, onboard processing satellites to meet the general needs of the business community.

Science Networks

ACTS provided real-time links to investigate the use of satellites for remote astronomy in two separate experiments. In the first user trial, T1-VSATs were installed at the Apache Point Observatory in southern New Mexico and on the campus of New Mexico State University (NMSU) in Albuquerque, New Mexico [76]. ACTS provided a 1.79 Mbps link to the observatory. The observatory's sensors produced a primarily digital output (e.g., a charged coupled device's image of objects in the sky), allowing the digital transmission of the observatory's data products in realtime. During an observation session, the telescope user remained at his home institution and did not have to be physically present at the telescope facility.

This link also allowed operation and control of the telescope facility by remote users, using the same interfaces as on-site observers. The ACTS trials allowed NMSU to test the capability of the remote interface, and give the user a touch and feel for access and control. In this aspect, the real-time nature of

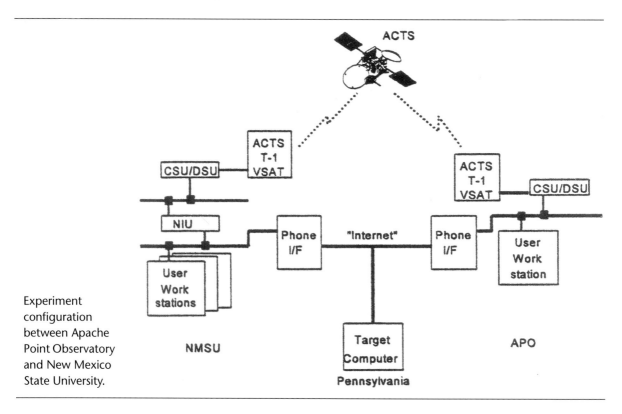

Experiment configuration between Apache Point Observatory and New Mexico State University.

the link was critical to the safety of the 11.5-foot telescope as it was moved under remote control. A wide band (1.544 Mbps) channel was also required to handle the data generated by the telescope's array of sensors. In addition, the ACTS trials included the testing and evaluation of additional communication links to support the non-real-time data network used to support observatory management, data base sharing, computer and video conferencing, and other collaborative services for the science community.

Overall, this experiment showed that the satellite link was a highly reliable means of delivering data to remote users from around the U.S. From a real-time control point of view, the ACTS links response time was very satisfactory and similar to that experienced in previous remote control tests using terrestrial links. The second trial involving the Keck Peak Observatory on the island of Hawaii is described in the next section on very high-data-rate applications.

Very High-Data-Rate Applications

The 900 MHz bandwidth of the ACTS transponders provided a unique capability to handle data rates of hundreds of megabits per second, which are not available with today's conventional satellites. HDR earth stations were devel-

T1-VSAT Installed at
the Apache Point
Observatory

oped to exploit this capability and deliver 155 Mbps (OC-3) and 622 Mbps
(OC-12) digital services. The use of SONET physical layer protocols allowed
seamless interconnections with the terrestrial fiber network. ATM communications were easily run on top of this SONET structure.

This capability opened up a whole new range of applications associated
with geographically distributed computing, especially those associated with
high-speed workstations and supercomputers. Although supercomputers were
once the domain of university and government researchers, additional non-
research applications are growing—such as the cancer dose treatment plan-

ning outlined in the section in this chapter on medical experiments. The satellite/terrestrial networking trials using ACTS provided the ability to make such supercomputers available on an on-demand, non-dedicated basis for those organizations and applications that don't require or can't afford a dedicated supercomputer. Such capability will provide more effective resource sharing and an improved utilization of computing resources.

Remote Computing

Aeronautical Modeling The Boeing Commercial Airplane Group in Seattle, Washington, conducted a series of interactive, computational flow, dynamic simulations to develop an engine control system by remotely flying an engine model in a numeric wind tunnel [77]. The inlet simulation that was developed by Boeing was executed on the LeRC Cray supercomputer and controlled by the Boeing engineers in Seattle. Flow visualization information from the Cray was transmitted via the ACTS gigabit network to Seattle, while setup and control information and commands were transmitted from Seattle.

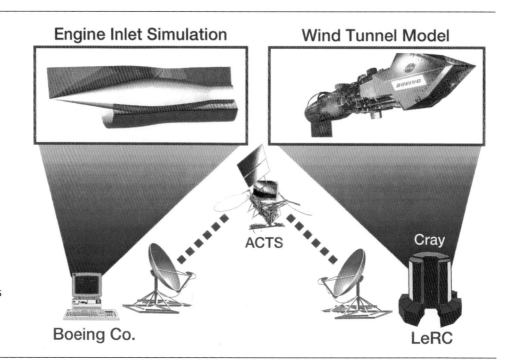

Boeing demonstrates high-speed data communications to reduce inlet control system design costs.

The large data throughput provided by ACTS allowed the Boeing staff to view flow visualization movies of the simulation in near realtime. The immediate feedback allowed the data to be examined and operation points to be readjusted in near realtime to develop the necessary inlet operating control characterizations. By comparison, Boeing (using local workstations) found that the iterative process of just determining good initial conditions would typically take several days and that running just one simulation would take weeks. ACTS made it possible to accomplish these processes in minutes.

Determining the results for multiple operating points required months on local workstations, while the ACTS-Cray combination streamlined the process down to a few hours. Not only did the ACTS interconnection physically speed things up, but it also enabled the quick detection of errant runs and allowed appropriate human intervention—permitting parameters to be quickly reset for another trial.

In addition, Boeing used some of the inlet characterizations gleaned from the ACTS tests to make more efficient use of the configurations being tested simultaneously in an actual wind tunnel. Boeing felt that in the future, use of a supercomputer via very high-data-rate connections could ultimately reduce resources, optimize wind tunnel tests (which are very expensive), and potentially alter design and manufacturing processes. In the future, companies may routinely access various facilities, such as supercomputers or wind tunnels, via high-speed links for more cost-effective utilization, higher productivity, and faster turn around.

While these inlet simulations were being run on the ACTS network, a team of Bellcore, Sterling Software, and LeRC engineers were evaluating and characterizing the performance of the ATM protocol being used to support the HDR transmissions. In general, the communication links performed very well, with zero ATM cell loss, a bit error rate comparable to fiber, and delivery of full raw-channel bandwidth. However the propagation time delay in going to and from the satellite can severely hamper the throughput of the communication channels. It was noted that the protocols and application codes needed to be modified to work well on a high-bandwidth, high latency network. The combined, computer-satellite links used a complex protocol stack of user datagram protocol (UDP), transmission control protocol (TCP), Internet protocol (IP), ATM, and SONET. Tests (with some modifications made to the protocols) demonstrated throughputs up to 58 Mbps for TCP and approximately 120 Mbps for UDP on a 155 Mbps ACTS channel. To improve these throughputs, further modifications needed to be made to the transport layer protocols. Such modifications were made and test results are reported in the upcoming section in this chapter on Standards and Protocols.

Astronomy A team from JPL, the California Institute of Technology (Caltech) and the University of Hawaii conducted a number of ACTS experiments between the Keck telescopes mounted on top of the Mauna Kea volcano on the island of Hawaii and an astronomy laboratory located at Caltech in Pasadena, California [78]. Astronomers are generally physically present at the observatory to conduct their investigations. In remote astronomy, however, an investigator collects a data set or image using a communication link between the astronomer's computer and the telescope. Networking instruments in this fashion allows more scientists to use the facility, permits more rapid analysis of the data, facilitates collaboration among the science teams, and even makes possible the use of the facility in a classroom setting. For this experiment, a combination of ACTS and terrestrial fiber (155 Mbps) links were used. In Hawaii, GTE fiber links connected the ACTS HDR terminal in Oahu to the Keck telescope, while in California the ACTS HDR terminal was interconnected to the CASA gigabit terrestrial fiber network at JPL. At JPL, a variety of supercomputers and high-speed workstations were used to enhance Keck observations in real time.

The high resolution of the telescope's various sensors generated images and data sets that were hundreds of megabits in size. Being able to process the data rapidly allowed faster and more extensive calibrations, as well as near real-time data examination. In the future, such high data communication links will reduce the time lost to travel and allow more scientists to use the facility and join in collaborative research.

Movie Production In the first steps to establish new, very high-data-rate global telecommunications networks, researchers in the United States and Japan cooperated in a multi-satellite/terrestrial hookup between the two countries to transmit high-definition video (HDV) [79]. The purpose of the experiment was to demonstrate broadband satellites' capability to deliver digital images at transmission rates up to 155 Mbps (OC-3). For this experiment, the Sony Studio in California was connected via terrestrial fiber to the HDR earth station at JPL. ACTS provided the link from JPL to Hawaii, where the data was then relayed via the Intelsat 701 Ku-band satellite to Japan and on to the Sony Studio via terrestrial fiber.

Sony's purpose in conducting this experiment was to investigate the rapid transfer of high-definition video (HDV) masters from remote shooting locations to post-production facilities for digital editing, dubbing, and the addition of special effects. In today's movies, the final image is frequently composed in post production to create the appearance of subjects being filmed in various backgrounds or amid special effects. In traditional filmmaking, it generally takes weeks and many iterations between the director and the film laboratory to produce images matching the director's vision. Long-distance collaboration between the people involved in the shooting and the editing will reduce processing time and expenses by allowing, among other things, scenes

136

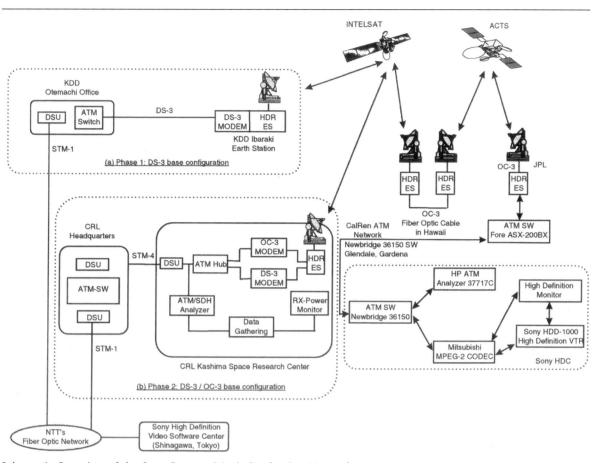

Schematic Overview of the Sony Remote Movie Production Network

to be re-shot in near real time while the actors and remote directors are on location and the set is still intact.

The projected business application of this experiment is the ability to interconnect all organizations involved in the making of a movie with global, wide-bandwidth links. The ultimate aim is to have complete electronic transmission of the image all the way from the studio to the theater. For this field trial, post-production compositing was successfully performed in Tokyo on a green-screen HDV clip transmitted from Los Angeles. This activity, along with the comparison of an HDV clip to its original source after one trans-Pacific satellite round trip, demonstrated the capability of the satellite channel to successfully transfer HDV. Once again, the striking capability of satellites to seamlessly interconnect with high-speed terrestrial networks at OC-3 data rates on a global scale was demonstrated.

Distributive High Speed Computing

Global Climate Modeling A JPL and GSFC team investigated global climate modeling using a coupled atmospheric-oceans climate model [78]. GSFC ran the atmospheric model, while JPL ran the ocean model. Boundary condition data was continuously exchanged via ACTS' 155 Mbps links, while these models were being separately processed on the JPL and GSFC supercomputers. A major objective was to explore approaches and performance issues for running complex models on heterogeneous computer architectures via a satellite link. With gigabit interconnections like ACTS, decomposition and distribution enabled the concurrent utilization of computational resources at geographically separated locations and sped up model execution. The models were successfully run using ACTS. The coupled, ocean-atmospheric modeling task is analogous to many of the future NASA modeling applications being planned in NASA's Earth Sciences Enterprise.

Great Lakes Weather Modeling A team consisting of the Ohio Supercomputer Center (OSC) in Columbus, Ohio; the National Center for Atmospheric Research (NCAR) in Boulder, Colorado; and the Great Lakes Environmental Research Laboratory (GLERL) in Ann Arbor, Michigan, performed a collaborative climate simulation in the Great Lakes region [80]. Cray supercomputers at OSC and NCAR, separated by 1,200 miles, were connected and operated in parallel via the ACTS 155 Mbps channels. GLERL was connected to OSC via a T3 terrestrial landline. All three sites interacted and collaborated while atmospheric, lake, and wave models were running in parallel at OSC and NCAR. This relationship required exchanging boundary conditions (i.e., surface heat and momentum fluxes, wave heights and direction, and lake surface temperatures). The results provided sophisticated flow visualizations of the air-lake interaction for Lake Erie for a given set of meteorological conditions. The success of this experiment illustrated how Great Lakes forecasters can generate better meteorological and marine forecasts using distributive computing.

As far as is known, the ACTS gigabit satellite network is the first one to provide connections at rates up to 622 Mbps, and with sufficient signal quality and transport mechanisms to be integrated with terrestrial fiber optical SONET networks. The ACTS gigabit satellite network provides BER performance comparable to fiber lines. As such, it demonstrates that satellites are a technically viable medium for providing very high data communication services alone or in concert with terrestrial networks as part of the information superhighway.

Oil Exploration

Oil exploration remains one of the most risky parts of the petroleum business. Most of the new frontiers of oil exploration are in deep waters, hundreds of miles offshore. Seismic acquisition vessels survey vast ocean areas, collecting data that can be analyzed by geophysicists hoping to locate the most promising drilling locations. A single ship can collect as much as 3 terabytes of raw data every week. Currently, data is collected on magnetic tapes and either flown from the ship or unloaded when the ship docks every few months. From these magnetic tapes, seismic data is transferred to a supercomputer center where sorting, screening, and other processes render it for study by oil company geophysicists. After review, the geophysicists may require additional data to be taken (which starts another long acquisition cycle involving a ship revisit), which may take six months to a year to arrange.

Delivery by satellite has the potential to revolutionize the process as well as the time frames in the collection and analysis of seismic data. The NASA ACTS team joined with the ARIES team to test and demonstrate the power of wideband satellite links, interconnected with multiple terrestrial networks, to preview the information superhighway of tomorrow [81, 82, 83]. ARIES, an acronym for ATM Research and Industrial Enterprise Study, was initiated at Amoco in 1993. The original goal was to build a small ATM, high-speed network with which to study the emerging ATM technology and how such networks could provide a competitive business advantage to Amoco. The initial 17 vendor partners that joined Amoco grew to 35, and the project has since been transferred to the American Petroleum Institute, where the widespread value to all members of the petroleum industry was recognized.

A series of application trials and demonstrations were conducted by the NASA-ARIES team to show that high-speed, satellite data links can extend the existing networks to remote and underserved regions not reached by fiber. A T1-VSAT earth station was placed on West Delta 90, an oil platform in the Gulf of Mexico, 40 miles off the Louisiana coast.

In another demonstration, the mechanically-steered, slotted waveguide antenna station (2 Mbps) was placed aboard the seismic exploration vessel, M/V Geco Diamond, which was plying the waters 120 miles offshore in the Gulf of Mexico. In these tests, the ACTS links demonstrated how megabytes of seismic or production data could be relayed in realtime to supercomputer centers for immediate analysis. The processed data was exchanged in near realtime with geophysicists at various oil company sites around the country over the ARIES terrestrial network. Collaborative interaction allowed additional processing, review, and analysis of the data sets. The use of ACTS compressed into days what normally took up to a year to complete. Researchers could literally steer

Part of the West Delta 90 platform in the Gulf of Mexico.

T1-VSAT mounted on the West Delta 90 platform.

140

the seismic vessel to reexamine promising areas before it left for other waters. Charles Dibona, president of the American Petroleum Institute, estimated that reduced time for completing seismic surveys could save the energy companies $200,000 a month for each survey—a considerable savings when you consider that hundreds of these surveys are conducted each year.

Oil Exploration Ship GECO Diamond

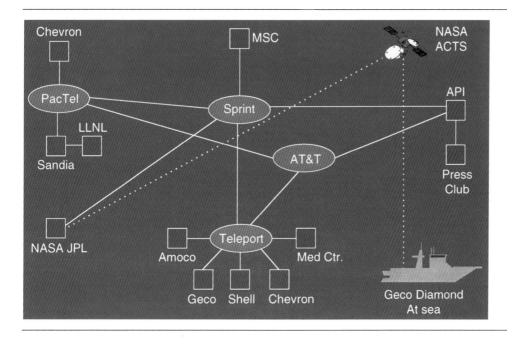

ARIES ACTS demonstration network involving the GECO Diamond.

141

ARIES ACTS team led by Dave Beering (center) of AMOCO.

The ARIES demonstrations also provided an excellent opportunity for the various application and network interface manufacturers and terrestrial wire-line carriers to test the interoperability of their various hardware, switching equipment, and fiber lines. The success of the ARIES/ACTS tests and demonstrations attest to the power of wide-bandwidth satellite links coupled with fiber to revolutionize businesses (like the petroleum industry) by compressing the time between data collection, processing, and analysis.

Global Interoperability

Videoconferencing

As has been discussed, ACTS has participated in a number of satellite/terrestrial interoperability trials and demonstrations. A satellite hop, a skip underneath the ocean on fiber, and finally a jump to the final destination via satellite describe the next demonstration in which ACTS was involved. On September 17, 1997, ACTS and Italsat-2 (the second operational Italian Ka-band satellite serving Italy and Europe), were used in the first two Ka-band satellite link connecting attendees at the third Ka-band Utilization Conference in Sorrento, Italy with participants in Canada and the United States. The network consisted of an ACTS link from Cleveland to the Canadian Research Centre in Ottawa, Canada, and a fiber link across the Atlantic and through Europe to the Telecom Italia CSELT laboratory in Turin, Italy. From Turin, the network was connected via Italsat-2 to Sorrento. Full duplex, 2.048 Mbps (E1, a European telephony standard analogous to T1) traffic was routed over the network using the ATM protocol. The live, multi-cast videoconference (at 128 Kbps video data rate per link) enabled sites in the United States, Canada, and Italy to interact simultaneously. Workstations using multiple windows displayed the speakers as well as the charts used in their presentations.

The overall quality of the video and audio was excellent. The content of the slides was clearly visible in the windows, while the small video of the speakers was enough to observe hand gesture, voice inflections, and general body motions. The delay of the double hop link did not detract from the videoconference. User application equipment was able to synchronize the video and audio to adjust for any delays. The double satellite transmission delay was detectable. Once the participants became accustomed to talking over this link, the impact of the delay became negligible. This demonstration provided another glimpse of the power of satellites working together with terrestrial networks to provide global interconnectivity. The demonstration could not have taken place without the tremendous cooperation of all the groups involved and the leadership of the Canadian Research Centre. It also showed the unifying power that common protocols could have in interconnecting totally separate and isolated systems.

Paul Mallasch at LeRC participates in the videoconference from Italy using desktop video monitors.

Aeronautical Applications

The demand for reliable data, voice, and video links between aircraft and ground has increased dramatically in the past decade. The clamor for increased passenger communication services is fueling this demand. In addition, reliable global aeronautical communication is required to support air traffic control, airline operations, and administration. The S- and L-band frequencies employed by the current systems support only low data rates (tens of

143

kilobits per second) and are becoming highly congested. Ka-band offers outstanding promise for accommodating aeronautical communication because of the large available bandwidth and higher data rates. In addition, Ka-band has the potential for supporting user equipment that is significantly smaller, and in some cases simpler, than L-band. Ka-band aeronautical links will not be plagued by shadowing and will largely avoid rain fades by flying at altitudes above the clouds. Satellite aeronautical communication should be an important new growth area for the telecommunication industry.

Low-Data-Rate Transmission Using MMIC Arrays

Aeronautical tests have been conducted via ACTS using two different types of antenna systems: proof-of-concept MMIC phased-arrays with electronic steering, and a slotted waveguide array with mechanical steering (see Chapter 3, "Terminal Equipment"). A NASA Lear jet was outfitted with proof-of-concept MMIC phased-array antennas built by Texas Instruments, Boeing, and Lockheed Martin [84]. Separate antennas at 20 and 30 GHz, with dimensions of only a few centimeters on a side, were used. Two-way voice and low-rate-data (at 4.8 and 9.6 Kbps) transmissions were successfully completed during the flights.

LeRC Lear jet with MMIC phased-array antennas mounted in the window.

This was the first time an electronically-steered, Ka-band MMIC array system had been demonstrated in an aeronautical terminal link with a satellite. As part of the test and evaluation program, demonstrations were provided to various airline, government, and equipment manufacturers. Demonstrations were held in both Cleveland and Dayton, Ohio; Washington, D.C.; Baltimore, Maryland; Boston, Massachusetts; Dallas, Texas; Los Angeles, California; and Seattle, Washington. The audiences at each location varied in composition, but included top-level representatives from American Airlines, Hughes, Mitre

Corporation, Lockheed Martin, Boeing, Texas Instruments, and Westinghouse. In addition, government representatives from the Air Force, Army, NASA, Rome Labs, and the National Communication System also attended. The successful results of the MMIC phased-array tests and demonstrations have created a strong incentive for continuing the focused development of MMIC array technology for satellite communication applications. In a letter to LeRC Director Donald Campbell, Larry Winslow (vice president of engineering for Boeing), cited that the development of phased-array antennas at Boeing for both commercial and military applications had been significantly aided by the LeRC ACTS demonstration program. He stated that "The performance demonstrated in these activities has contributed to the overall concept credibility, adding assurances that the continued development will have a worthwhile payoff." He went on to state that "NASA's work, which is designed to further development of technology that can be used by American industry, has often accelerated the rate of progress possible by the industry." In the future, the use of small, conformal, phased-array antennas will provide electronic steering, lower installation costs, and lower drag profiles than mechanically-steered reflectors.

Slotted Waveguide Antenna at T1 Rates

Using the slotted waveguide antenna, two different series of aeronautical tests and demonstrations were conducted [85]. In the first, the antenna was mounted in a Rockwell Collins Saberliner 50 aircraft. Serving both the commercial and government aeronautical markets, Rockwell was interested in validating and characterizing the T1 transmission performance of a Ka-band system to and from a business aircraft with an eye toward future business development.

In the second test, the antenna was mounted and flown aboard NASA's Ames Kuiper Airborne Observatory (KAO). The KAO (a C-141A jet aircraft), carries a 0.9 meter reflecting telescope used for infrared astronomy. With the KAO, a number of successful experiments were carried out, including the PBS-sponsored "Live from the Stratosphere."

"Live from the Stratosphere" was a multimedia, educational program that included live, two-way video, audio, and Internet-type transmissions, as well as broadcasts from KAO to the ground. The video and audio were sent via a combination of commercial satellite and landlines to Public Broadcast Service (PBS) stations on the East Coast, where the program was produced and sent out live for broadcast on PBS channels and NASA TV. During the broadcasts, selected sites at museums and schools were given the opportunity to communicate directly with the crew and scientists aboard the KAO. One group of students at the Adler Planetarium in Chicago was able to control the KAO telescope using an Internet link. During other KAO research flights, scientists from the University of Chicago's Yerkes Observatory tested and demonstrated remote control and operation of the telescope aboard the aircraft. The scientists had immediate access to the telescope's imagery, and CUSeeMe teleconferencing over the link allowed real-time scientific collaboration. This experiment showed that interactive, videoconferencing sessions could be successfully accomplished with two satellite hops. During these research flights, another group successfully tested and evaluated a remote failure- and diagnosis-monitoring system using the Internet. Such a capability eliminates the need for a technician to travel on the aircraft.

The aeronautical trials pioneered and tested using ACTS, served as a proof-of-concept for global, wideband, aeronautical communication services.

Maritime Applications

The use of the slotted, waveguide antenna was already chronicled in the petroleum exploration tests and demonstration with the seismic vessel, GECO Diamond. This mobile terminal was also installed on the U.S. Navy cruiser U.S.S. Princeton (CG 59) and remained aboard for one year—during which time the Princeton was deployed in the Eastern Pacific, Sea of Cortez, and the Caribbean [86]. The U.S. Navy wanted to increase the capacity of the communication links available to its deployed forces and looked to ACTS to provide a chance to test an early prototype of future operational capacity. The ACTS terminal provided full-duplex communication at data rate up to T1 (1.544 Mbps).

The link was used for voice, Internet, and video traffic. These tests also provided the first Ka-band mobile propagation data in a marine environment. The ACTS link gave the U.S.S. Princeton access to medical, logistical, maintenance operations, training, personnel, and weather-forecasting services. In addition to directly enhancing the Princeton's mission effectiveness with work related services, ACTS allowed officers and crew to stay in touch with their families via interconnections through the public telephone network and Email.

While cruising off the coast of southern California in February of 1997, the Princeton received word that the 66-year-old master of the Greek bulk cargo vessel, Phaethon, was in medical distress with severe abdominal pains and cramping. The Princeton proceeded several hundred miles to rendezvous with the Phaethon. The Princeton's medical officer rendered immediate aid and used real-time, videoconferencing via ACTS with the Naval Medical Center in Balboa, California. A urologist from Balboa, examining the patient via the ACTS video link, determined that he had a blocked urinary tract and guided the Princeton's medical officer through a specialized course of treatment. The opportunity for timely intervention that was provided by the ACTS link was later recognized by the Navy and the doctors providing follow-up treatment as essential to saving the life of the Greek captain.

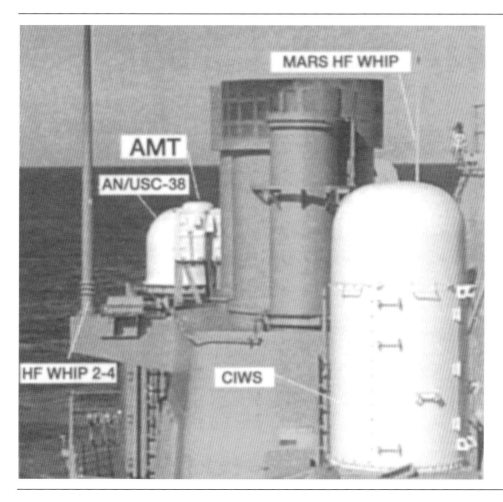

The ACTS mobile terminal (AMT) on the U.S.S. Princeton as viewed from the starboard bridge wing, looking aft.

The communication services that ACTS brought to the Princeton served as a model for the future of satellite communication services on all Navy ships. With extended deployment, a ship's need for greater connectivity with the information infrastructure ashore becomes critical for mission effectiveness and morale. ACTS provided the Navy with invaluable opportunities to demonstrate the many ways that full-duplex, high-data-rate satellite communication can enhance the quality of life at sea.

In a continuing effort to enhance satellite communication connectivity to naval forces, the Office of Naval Research once again teamed with NASA. In

October of 1998, LeRC, the Naval Research Laboratory (NRL), and a team of various commercial equipment suppliers successfully demonstrated a two-way mobile network operating at 45 Mbps. LeRC engineers assembled a 4-foot diameter terminal from commercially available hardware. Using a 120-watt TWTA and a fully articulated tracking pedestal, they transferred data from the Motor Yacht Entropy that was operating in the choppy waters of Lake Michigan, via ACTS, to the HDR terminal at LeRC. Communication transfers included TCP/IP-based file transfers, interactive and variable TCP/IP-based multimedia, production-quality video, and CD-quality audio.

At Sea Testing of the HDR Terminal

The 45-foot Entropy with 4-foot antenna for providing two-way 45 Mbps data communications.

Supervisory Control and Data Acquisition Applications

The use of Ka-band and spot beam antennas has the potential to reduce the size and cost of earth stations for low-data-rate (Kbps) applications. Ultra Small Aperture Terminals (USAT) with an antenna diameter of 14 inches and less than one watt of transmitting power can be used to provide supervisory control and data acquisition (SCADA) services. SCADA systems typically involve a central processing point collecting real-time information on a continual basis from remote points for purposes of monitoring and control. Systems of this type are employed in electric, gas and water utilities, oil and gas production fields, and gas pipeline systems.

SCADA applications are one example of a much broader class of low-throughput data applications. This general class of applications represents a very large market that could be cost-effectively served with the type of low-cost ground terminals made possible by the ACTS Ka-band spot beam technologies.

In a joint undertaking, NASA LeRC and Southern California Edison (SCE) tested the 14-inch diameter USAT at data rates of 4.8 Kbps, in the SCE operational SCADA network in southern California [87]. The purpose of this network is to monitor the status of the electrical distribution grid and provide path and redundancy switching for failure recovery. In this test, a SCADA master computer at SCE's Vincent Regional Control Center polled a USAT at the unmanned, Goldtown substation every 4 seconds. The availability of the Ka-band system was compared with that of the Ku-band VSAT system currently in use by SCE.

Managers from Southern California Edison and LeRC meet to discuss the plans for the upcoming SCADA experiment in California.

SCE alone has over 10,000 points in their electrical grid over a 50,000 square mile area which they would like to monitor and control. Such wide geographical dispersal provides an ideal stage for the use of satellites. At the time of this experiment, they were only monitoring approximately 900 major substations because of the relatively high price using Ku-band VSATs. The price objective was $75 per month per location. The hope was that, with the development of Ka-band commercial systems, the price for conducting SCADA operations would become inexpensive enough to allow widespread deployment of a SCADA satellite network.

Distance Education Applications

Education and training are key to increased personal knowledge and national productivity. In this era of changes in the classroom and evolution in the workplace, students and employees need better access to educational and

149

training facilities to remain professionally competent. Satellites have been, and continue to be, used to deliver broadcast-type distance education and training. In 1996, more than 70% of corporate satellite networks were used for training and new product information. Distance learning offers a dynamic method to train hundreds or thousands of employees simultaneously.

Part of the Georgetown University Team: (left to right) Rev. Harold Bradley; Dr. Irma Frank; Carlos Lietas, Columbian Ambassador; Rev. Byron Collins; and Dr. Fred Ricci.

There are about 6.9 million students attending some 22,400 rural schools in the U.S. This accounts for 17% of the regular public school students and 28% of regular public schools. Nearly three-quarters of all rural public schools have fewer than 400 students. To serve these smaller enrollments, innovative instruction practices are needed to educate these students and train their teachers adequately. Satellite communication, seamlessly integrated with the terrestrial network, will contribute significantly to this process. High-data–rate, low-cost communications, using small earth stations offered by systems like ACTS, have the potential to raise the use of satellite-delivered education to a new plateau.

For the ACTS distance learning trials, each duplex video link was at a minimum of 384 Kbps (maximum of T1), and Compression Laboratories video equipment specifically designed for distance learning was installed at all locations. The multiple channels offered by the T1-VSATs allow video broadcasts of classroom instructions as well as the simultaneous capability for return video and audio links to allow better instructor/student interaction.

Three different teams used ACTS to investigate various aspects of distant education. Georgetown University (GU) led the Latin America Distant Education Experiment. GU conducted a number of different types of classes with

Javeriana University in Bogota, Colombia, and Catolica University in Quito, Ecuador.

Working with a team of private industry organizations and the World Bank, GU evaluated the effectiveness and economic viability of offering not only academic courses, but small business development training classes as well. In developing countries, satellites offer the most reliable and cost-effective means to conduct overseas training programs. A location can be immediately connected into a satellite network without reliance on local terrestrial links.

GU offered classes and seminars in business, Latin American studies, medicine, nursing, and linguistics [88, 89]. Seminars on small business development were targeted to small enterprises that operate in Colombia and Ecuador. The seminars covered such topics as marketing, growth and expansion, import/export regulations, and capitalization. The intent of the program was to increase the resource base for small businesses in Latin America. A year-long series of classes were conducted at GU, where students in Latin America attended via ACTS. The transmission quality was excellent, and the system provided an opportunity not only for the University and Small Business Development classes, but also a means for conducting a number of telemedical conferences.

Georgetown University holds Latin American Small Business Seminar with Javeriana University in Bogota, Columbia, using ACTS.

The ACTS satellite education program enhanced the regular classroom experiences of the GU students by incorporating professor and students from Latin America into the classroom via satellite. Their experience shows that dis-

151

tance education has the potential to dramatically change the teaching of international studies by incorporating foreign students and scholars directly into the curriculum.

ACTS provided for more interactions between the instructor and the student by moving distance education learners out of a one-way classroom environment and into a more traditional one. As one author put it, "Technology is taking the distance out of distance education." With the potential for Ka-band satellites to deliver the necessary service for distance learning at a low cost, this superior capability will become much more affordable.

Standards & Protocols

One of the key challenges facing the satellite industry today falls under the banner of standards and protocols. Until recently, satellite networks have typically been operated as isolated, private networks. If satellites are to play a key role in the NII/GII, however, they must provide seamless interoperability with both existing and evolving terrestrial networks. There are terrestrial data protocols and standards in use today (or under development) that are either completely incompatible with satellite systems or make a satellite network very inefficient. This is especially true for ATM and the Internet protocol suite TCP/IP—the two main protocol standards being implemented in high-speed terrestrial networks.

Many companies and organizations have become involved in research and development of TCP/IP over satellite links. These include Bellcore, COMSAT, Lockheed Martin, Hughe, Motorola, Cisco, Orion, Cray, Sterling Software, Mitre, NCAR, OSC, Ohio State University, Ohio University, BBN, University of Maryland, Ames, ATT, Teledesic, LeRC, GSFC, and JPL [see references 90-99].

In June of 1998, the NASA LeRC held a major conference in Cleveland, Ohio, that was dedicated to satellite protocol issues, in which much of the current efforts were highlighted. NASA was urged by the attendees to continue using ACTS to promote advances in satellite/terrestrial interoperability [100]. As a result of this conference, numerous additional approaches to addressing the effects of satellite channel errors and latency have been investigated using ACTS. The following are a few of the many ACTS accomplishments:

- Ohio University developed a modification to TCP/IP that allowed file transfer to take place at nearly the full 1.5 Mbps rates of the T1-VSATs.

- Bellcore conducted a series of ACTS experiments to test and evaluate seamless operations for delivery of personal communication services—especially wireless packet—to scope out issues that may be important at the L- and S-bands for mobile services.

- In tests using the ACTS gigabit network, NCAR and OSC modified and enhanced the TCP/IP and achieved transfer rates over 120 Mbps in a 155 Mbps OC-3 channel.

One of the most important protocol experiments carried out over ACTS has been dubbed 118X [101]. This indicates one satellite, 18 organizations, and the dispelling (X-ing) of the myth that the delay inherent in GEO satellite transmission precludes operation of the TCP protocol. The ACTS high-data-rate team has been working with industry and other government groups (shown in the table below), to test and demonstrate the performance of TCP/IP over ACTS in an effort to dispel the myths about operating with TCP/IP over very high speed satellite links. In late 1998, this team, consisting of computer, networking, and satellite companies, as well as government laboratories, tested key TCP/IP performance enhancements. Their primary goal was to optimize the point-to-point transfer of data between two locations across the satellite, using TCP/IP over ATM. The memory-to-memory transfer tests over the ACTS 622 Mbps ATM link were conducted using a wide variety of computers and operating systems. Throughput performances were achieved in both homogeneous (over 500 Mbps) and heterogeneous (over 350 Mbps) vendor environments. Such successful trailblazing experiments and demonstrations help to hasten the introduction of such services in future generations of geo-stationary communication satellites.

The Joint Industry-Government TCP/IP High-Speed Performance Experiment Team

Computer Industry	Communications Industry	Satellite Industry	Government Laboratories
Sun	Sprint Advanced Technology Laboratory	Lockheed Martin	NASA Lewis Research Center
Microsoft	Cisco Systems	Hughes Space & Communications	NASA Johnson Space Center
DEC	Fore Systems	Space Systems Loral/ Globalstar	NASA Jet Propulsion Laboratory
Intel	Ampex Data Systems	Spectrum Astro	Lawrence Livermore National Laboratory
Pittsburgh Super-computing Center			Naval Research Laboratory

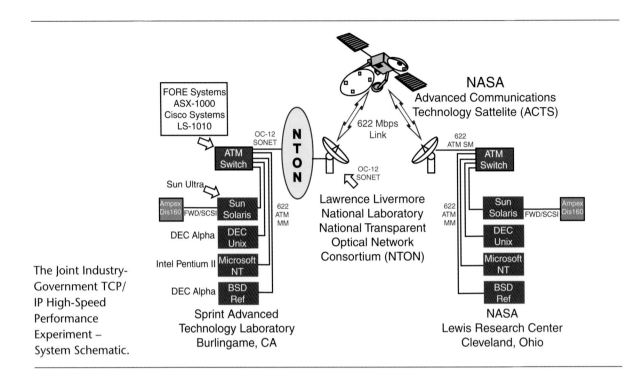

FORE Systems
ASX-1000
Cisco Systems
LS-1010

OC-12
SONET

NTON

ATM
Switch

Sun Ultra

Ampex
Dis160

FWD/SCSI

Sun
Solaris

DEC Alpha

DEC
Unix

Intel Pentium II

Microsoft
NT

DEC Alpha

BSD
Ref

622
ATM
MM

NASA
Advanced Communications
Technology Sattelite (ACTS)

622 Mbps
Link

OC-12
SONET

Lawrence Livermore
National Laboratory
National Transparent
Optical Network
Consortium (NTON)

622
ATM SM

ATM
Switch

622
ATM
MM

Sun
Solaris

FWD/SCSI

Ampex
Dis160

DEC
Unix

Microsoft
NT

BSD
Ref

Sprint Advanced
Technology Laboratory
Burlingame, CA

NASA
Lewis Research Center
Cleveland, Ohio

The Joint Industry-Government TCP/IP High-Speed Performance Experiment – System Schematic.

Concluding Remarks

When the foundations of the ACTS program were being debated and established in the early 1980s, the first personal computers were just being introduced by IBM. Today we have a plethora of communication devices—not only at work, but also in our homes. Personal computers, faxes, Email, cordless and mobile phones, pagers, and the Web are almost indispensable in the fabric of our everyday business and personal lifestyles. Access to information is essential for national and international economic development.

ACTS triggered and accelerated the shift from the static, bent-pipe satellite repeaters of the past to a new stable of agile, on-demand, wide bandwidth, all-digital class of satellites that will lower the price for interactive services. In the 1980s, satellites served primarily top U.S. corporations and the government. The forthcoming generation of Ka-band satellites, modeled after ACTS, will provide services to not only a whole range of businesses, but potentially into the home as well.

Through the ACTS user program, the basic technologies of switching, spot beam satellites have been tested, characterized, and validated. Trials with the ACTS networks have demonstrated the use of digital, on-demand, integrated services via satellites and shown that Ka-band spot beam systems with onboard switching and processing can provide these services reliably and with suitable availability.

CHAPTER 5

KA-BAND PROPAGATION EFFECTS

Knowledge of signal propagation characteristics is of fundamental importance to the design of communication satellites and their associated user-terminal networks [102, 103]. The relative importance of the various propagation factors depends to a large extent on the radio frequency of the transmission, the local climate, and the elevation angle to the satellite. Generally, the effects become more significant as the radio frequency increases.

In the design of Ka-band communication satellite systems, a critical requirement is achieving satisfactory service availability. Propagation effects, which can seriously degrade the reliability and performance of a communication link, strongly influence the choice of suitable satellite system parameters.

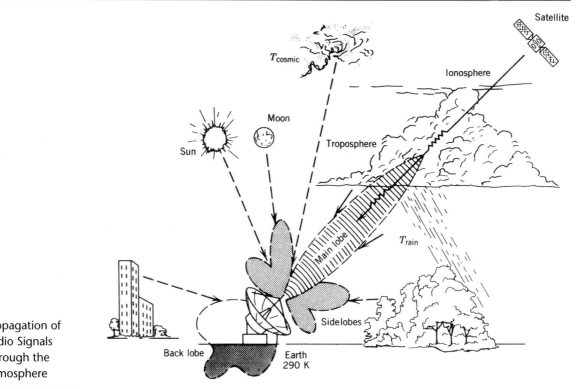

Propagation of Radio Signals Through the Atmosphere

Rain is the primary factor affecting satellite links. At the 20 and 30 GHz frequencies of the ACTS system, rain attenuation is a more serious problem than at the lower C- (4 and 6 GHz) and Ku- (12 and 14 GHz) band frequencies. For example, for a rain rate of 22 millimeters per hour in Washington, D.C., the rain attenuation for C-band is 0.1 dB/km, Ku-band is 1 dB/km, and Ka-band is 5 dB/km.

Physically, rain attenuation is due to scattering and absorption of the microwave signal energy by the raindrops. As the radio frequency increases,

156

the wavelength decreases and the signal attenuation in rain increases. For instance, at 30 GHz, the wavelength is 1 cm compared to a typical raindrop, which is 0.15 cm in diameter. The droplets act as lenses, bending the waves and distorting the signals. As the rain rate increases during a heavy downpour, the size and number of the raindrops increases, and so does the attenuation. It is the rain rate, not the yearly rainfall amount, that determines the link availability.

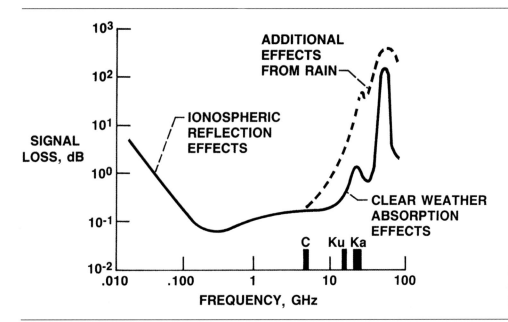

Signal Attenuation as a Function of Frequency

Although rain is the dominant propagation impairment of Ka-band signals, second order impairments due to gaseous absorption, cloud attenuation, tropospheric scintillation, and antenna wetting can become significant.

Gaseous absorption is the reduction in signal strength due to absorption by the gaseous constituents of the atmosphere—primarily oxygen and water vapor—which are present in the signal path to and from the satellite. Losses can be up to 1 dB. Clouds and fog, which consist of water droplets of less than 0.1 mm in diameter, can also attenuate the Ka-band signal on the order of 1 dB.

Scintillation refers to the rapid fluctuations of the radio signal's amplitude and phase, on a scale of a few seconds to tens of seconds, which are associated with fluctuations in the index of refraction produced by turbulence. Rapid fluctuations in the troposphere, caused by high humidity and temperature gradients, change the refractive index along the propagation path. The effects are seasonally dependent, vary day-to-day, and vary with the local climate. Scintillation is more significant for earth terminals with low elevation angles to a geostationary satellite like ACTS, from locations like Alaska.

Raindrops, ice crystals, and hailstones can cause the depolarization of the signal from one polarization state to another. Many satellite communications systems use dual orthogonally polarized transmissions to achieve frequency reuse or to increase the isolation between adjacent beams or between nearly collocated satellites that use the same frequencies. For these systems, the resulting interference must be accommodated.

Absorptive path losses are accompanied by increased sky brightness (radio noise), which appears to a ground-based receiver looking at the satellite as increased background noise. This results in a decrease in receiver sensitivity. At 20 GHz, the degradation can be as great as 40% of that caused by the path attenuation.

Rain, as well as snow, can also cause additional signal losses by wetting the antenna and/or the surfaces of the feed horn. The ACTS experimenters have studied this effect, which appears identical to a path loss but is very difficult to identify separately and therefore to model.

Reliable designs of Ka-band systems require not only accurate estimates of the individual impairments, but also an estimate of the statistics of their combined effects. Data collected from the ACTS beacon receivers provides a more reliable basis from which to understand these effects and develop models and predictive procedures.

Link Margins

Propagation impairments, acting alone or in combination, can cause severe degradations in the signals to and from a satellite. For digital systems, like ACTS, these degradations will increase the bit error rate.

The traditional method of accounting for the effects of atmospheric impairments on satellite links is the assignment of link power margins to provide acceptable service performance. As discussed above, the performance of Ka-band systems will be affected by atmospheric impairments to a greater degree than for C- or Ku-band systems and must include the effects of gaseous attenuation, cloud attenuation, and tropospheric scintillation, in addition to the rain attenuation. The ACTS propagation measurement campaign was designed to supplement previous measurements for all these effects and provide a more accurate prediction capability for the rain zones of North America. A more accurate prediction capability is especially needed for those systems providing services with low availability (low link margins). NASA/JPL contracted with Stanford Telecommunications to revise the NASA handbook on calculating link margins in the 10 to 100 GHz frequency range. This revised handbook [104] provides the latest methods for calculating the necessary link margins to achieve the desired availability.

Ka-band Propagation Statistics

The problem of predicting RF propagation attenuation is greatly complicated by its statistical (random) nature. At any one location, the rain characteristics will vary greatly from year to year, although recent findings indicate that a seven-year cycle provides a very reliable indication of long-term effects. As a result, measurements need to be taken over many years to determine a valid statistical average. Statistics of the percentage of time that a predicted level of degradation is exceeded are among the propagation quantities needed by researchers. Information on impairment details, such as the average duration of an impairment level, may also be required—especially for the design of adaptive impairment-mitigation systems, design of user terminals, and sizing of satellite resource-sharing methods.

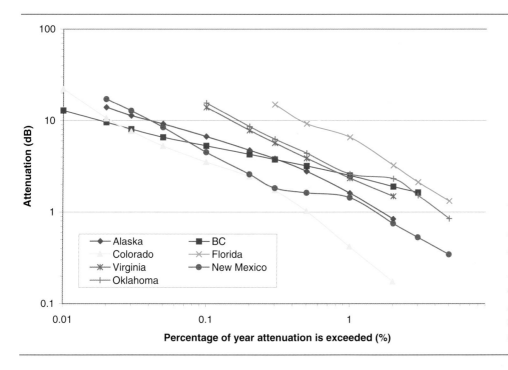

Five-year ACTS average annual rain attenuation at 27.5 GHz for seven sites. The measurement period is for 1994–1998 [105]. Attenuation is with respect to clear sky corrected for wet antenna effect.

Data provided courtesy of Dr. Robert K. Crane of the University of Oklahoma.

Since the 1970s, a number of satellites have carried Ka-band beacons to characterize propagation effects, especially those caused by rain [63]. These have included the NASA application technology satellites (ATS-5 and ATS-6), COMSTAR (USA), ETS-II (Japan), SIRIO and Italsat (Italy), and Olympus (ESA), and CS (Japan). All but two of the propagation measurements made within the United States, using satellite beacons, were obtained in a single rain climate zone. The rain climate varies widely within the United States, and additional observations and measurements were clearly needed in those climate regions where no Ka-band data existed. As a result, the ACTS program decided to include beacons on the satellite and to sponsor measurements in a variety of rain zones.

NASA Propagation Campaign

The NASA Propagation Experiment group (NAPEX) has been in existence since the 1960s, sponsoring various experimental and theoretical studies, and model developments on the characterization and effects of earth-space path radio-signal impairments at various frequencies. NAPEX has always had a strong partnership with private industry, other government agencies, and universities to leverage NASA assets and other resources for obtaining propagation data. The NAPEX group first discussed redirecting its efforts to concentrate on recording and collecting Ka-band propagation measurements and studying fade mitigation techniques [106] in support of ACTS at a 1987 meeting at COMSAT.

In November of 1989, NASA (through NAPEX), formally announced its decision to support ACTS propagation measurements and began the planning for such measurements by sponsoring the first ACTS Propagation Studies workshop (APSW) in Santa Monica, California. Representatives attended this workshop from private industry, academia, NASA, and other users of propagation data. At this workshop, the group addressed a number of topics, including the need for propagation data and the configuration and number of propagation terminals required for gathering the data. Participants defined the overall goals of the ACTS propagation campaign and delivered a set of recommendations regarding propagation studies and experiments that would utilize ACTS. They also provided guidelines regarding measurement parameters and requirements. In subsequent workshops, held at the rate of two a year, the plan was refined and top-level requirements were generated for the development of a receive-only propagation terminal and the data collection software.

The overall goals of the ACTS propagation campaign were to:

- characterize all the important Ka-band frequency impairments, including attenuation due to rain, clouds, gaseous absorption, and scintillation

- determine the characteristics of fade rate and duration on Ka-band satellite links

- expand the propagation research base in all United States rain climate zones

- study fade compensation techniques, and

Bob Bauer (left), NASA LeRC, and Dr. Faramaz Davarian (right), NASA JPL—coordinators of the ACTS propagation program. After he left JPL, Dr. Davarian was succeeded by Nasser Golshan.

- characterize Ka-band propagation effects on mobile satellite links

Although most mobile and personal communications systems will operate below 3 GHz, frequency congestion will inevitably force system planners to use the Ka-band frequency for such applications as aeronautical and maritime mobile communication. In addition, the military has a strong interest in using Ka-band for mobile communication. At the time of the ACTS program, no method was available to make Ka-band satellite link predictions for mobile applications.

ACTS Propagation Experiments

The ACTS program released a NASA research announcement (NRA) entitled "ACTS Propagation Experiments Implementation Program" for funding propagation measurements [62]. In May of 1992, ten organizations were selected to carry out a series of propagation measurements.

Long-Term Attenuation Measurements

Seven experiments involving ten organizations were selected to receive the ACTS propagation terminals for making long-term, in situ measurements.

SITE	ORGANIZATION	PRINCIPAL INVESTIGATOR
Clarksburg, MD	COMSAT	Asoka Dissanayake
Fairbanks, AK	University of Alaska	Charles Mayer
Ft. Collins, CO	Colorado State University	John Beaver
Las Cruces, NM	Stanford Telecommunications New Mexico State University	Louis Ippolito Stephen Horan
Norman, OK	University of Oklahoma	Robert Crane
Tampa, FL	Florida Atlantic University University of South Florida	Henry Helmken Rudolf Henning
Vancouver, BC	University of British Columbia Communications Research Centre	C. Amaya David Rogers

These selected sites span many rain climate zones from subarctic Alaska to subtropical Florida, and from arid New Mexico to temperate Maryland. The slant path to ACTS encompassed a range of elevation angles from as low as 8° for Alaska to as high as 52° for Florida. NASA funded experimenters at these sites to collect and analyze data for five years. The five-year period was chosen for the following reasons:

1. Attenuation by rain is a random process. The percentage of time exceeded versus fade depth distributions observed in one year may be a poor predictor of the distribution observed in the following year. With only two years of data collection, statistical errors will far exceed measurement errors.

ITU-R Rain Zones & Rainfall Rate Exceeded (mm/h)

% of Time	C	D	E	K	M	N
1.0	0.7	2.1	0.6	1.5	4	5
0.1	5	8	6	12	22	35

ACTS Propagation Experiment Sites (ITU-R Rain Zones)

2. Past empirical observations have shown weather cycles lasting seven years. For this reason a five-year observation period was determined to be the minimum duration.

3. Originally, the plan was to gather data for only two years. However, after two years, some of the sites showed results that substantially deviated from the norm.

4. A subgroup of the Satellite Industry Task Force (SITF) strongly recommended that the program be extended to five years.

163

Other Propagation Experiments

Three other experiments (involving four organizations) were selected to conduct shorter-term investigations related to other aspects of propagation. These experiments included wide-area diversity, uplink power control, land-mobile satellite measurements, and satellite-channel characterization. In addition, LeRC investigated fade mitigation techniques using adaptive convolutional coding and burst rate reduction, wet antenna effects, depolarization effects, and collected propagation data at the master control station in Cleveland. JPL made propagation measurements as part of the various land and aeronautical mobile communications tests and experiments they conducted.

EXPERIMENT	ORGANIZATION	PRINCIPAL INVESTIGATOR
Uplink Power Control Wide Area Diversity	COMSAT	Asoka Dissanayake
Land Mobile Satellite Measurements	Johns Hopkins University University of Texas	Julius Goldhirsh Wolfhard Vogel
Propagation Effects on Digital Transmissions	Georgia Tech Research Institute Georgia Institute of Technology	Daniel Howard Paul Steffes

Propagation Terminal Development

The development of a receive-only ACTS propagation terminal (APT) was initiated in parallel with the solicitation and selection of propagation experiments. The guidelines developed for the ACTS propagation campaign recommended that identical receive-only APTs be built so that Ka-band propagation data could be collected and processed in a uniform and consistent manner [64]. A common terminal allowed for more accurate site-to-site data comparisons and a common support and maintenance effort.

A fortuitous test bed opportunity to experiment with Ka-band beacons prior to developing a prototype ACTS propagation terminal was provided in 1989, when the European Space Agency (ESA) launched its telecommunications satellite, Olympus. Olympus carried a number of propagation beacons that transmitted stable signals at 12.5, 19.7, and 29.7 GHz, all of which were coherently derived from a common source. NASA (through NAPEX) funded the Satellite Communications group at Virginia Polytechnic Institute (VPI) in Blacksburg, Virginia, and the Westenhaver Wizard Works to develop a propagation terminal for use with the Olympus spacecraft. The Olympus test bed resolved technical, data-collection, and analysis problems. This terminal design was the basis for the ACTS propagation terminal.

The ACTS propagation terminal requirements included:

- a beacon receiver dynamic range of greater than 20 dB for continental U.S. locations,

- a measurement accuracy of 0.5 dB, with a resolution of 0.1 dB, and

- a sampling rate of 1 Hz to provide the necessary rain attenuation time series

A second sampling rate of 20 Hz was chosen for short-term measurement of scintillation events. The APT consisted of a common 4-foot antenna, a dual-channel, digital receiver, a dual-channel radiometer, and a data acquisition system for making measurements at both 20.2 and 27.5 GHz. It was also equipped with meteorological recorders for measuring the point rain rate and the atmospheric temperature and humidity. A major advantage of the digital receiver was that it acquired the signal in less than 3 seconds. In addition, if a signal was lost in a deep fade, it would be reacquired as soon as the attenuation became less than 25 dB.

In order to provide valid probability distribution estimates from long-term propagation observations, a reasonably complete data set must be collected. This required, in turn, that the equipment be in operation for a large percentage of the time. The APT specification was set at a better than 90% equipment availability and independence between equipment failures and rain occurrences.

Spacecraft Beacons

The ACTS propagation payload consisted of a pair of vertically polarized 20.2 and 27.5 GHz beacons, known as downlink and uplink beacons, respectively. Each beacon was backed up with a spare. The spare for the 20.2 GHz beacon was horizontally polarized. The uplink beacon was pure CW transmission, whereas ranging tones and low bit-rate telemetry modulated the downlink beacon. The design of the downlink beacon, however, incorporated placeholder tones to prevent a change in the carrier's power level when the ranging signals were turned off. The beacon radiation pattern covered the contiguous U.S., most of Alaska, Canada, and northern Mexico. The effective isotropic radiated power (EIRP) values of the 20.2 and 27.5 GHz beacons were approximately 17.6 and 15.9 dBW, respectively, at the edge of CONUS coverage.

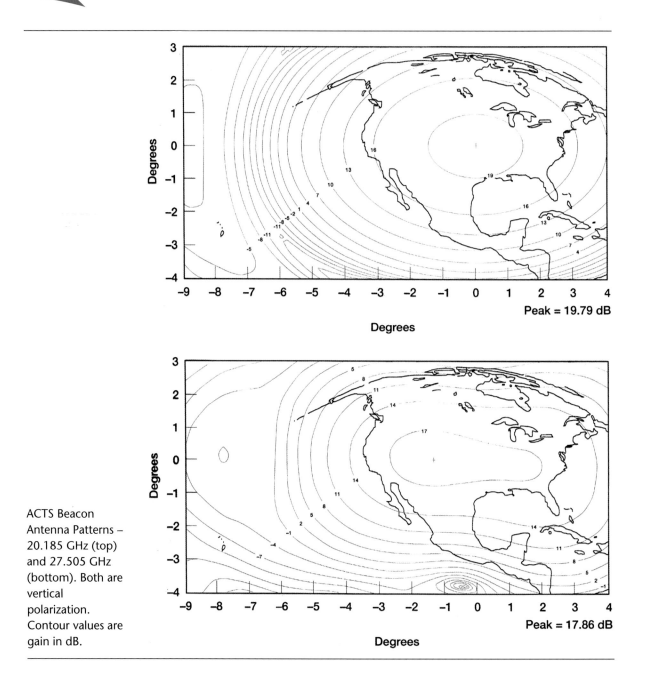

ACTS Beacon Antenna Patterns – 20.185 GHz (top) and 27.505 GHz (bottom). Both are vertical polarization. Contour values are gain in dB.

Propagation Experiment Operations

All propagation terminals were deployed and ready to collect data by the time ACTS was ready for launch. On Thursday morning, September 23, 1993, at approximately 10 a.m. EDT, the 20.2 and 27.5 GHz beacons were switched on for the first time while ACTS was drifting to its 100° west longitude location.

Dr. Charles Mayer, University of Alaska, explains the low elevation angle of the Alaska ACTS propagation terminal.

The University of British Columbia acquired the beacons at 11:02 a.m. EDT. Over the next day and a half, all the other propagation sites acquired the beacon. The University of Alaska acquired amid a few snowflakes, and the University of New Mexico captured data from a rain event on the first day that it received the beacon signals. The time period from launch through the end of

167

November was used as a shakedown period to learn how to effectively use the propagation terminals and wring out the system. An improved version of the preprocessing software was completed and delivered to the experimenters. On December 1, 1993—after completing all checkouts—the propagation experiments were declared operational and the data collection phase began.

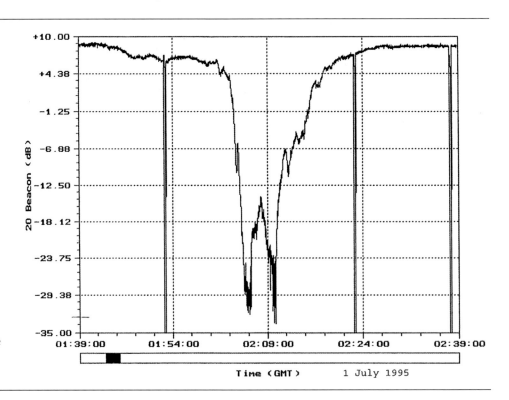

Rain fade event at White Sands, New Mexico, July 1, 1995 (fades much greater than 25 dB are suppressed due to limitations in dynamic range).

The basic recording mode of the APT collected beacon and radiometer data continuously at one sample per second. This data was combined with a time stamp, meteorological observations, and recorder status information, and stored in daily output files. These raw or unprocessed daily data files were preprocessed to produce attenuation, fade duration, interfade intervals, sky brightness, temperature, and rain rate histograms.

While beacon data collection continued, other propagation experiments were also being conducted. The University of Texas and the Applied Physics Laboratory of Johns Hopkins University collaborated on a mobile propagation investigation [57]. The objective of these measurements was to establish the extent of fading caused by roadside trees, obstacles, terrain, and urban structures on the 20 GHz downlink signal emanating from ACTS. Data, which was collected using a van with a computer-controlled tracking antenna mounted on the roof, will be used to construct models for use in future mobile communication satellite designs. Measurements were taken in Texas with bare tree conditions, and in Maryland and Alaska with full foliage trees. The Maryland measurements were made over the same set of roads previously used to collect data at UHF and L-band, enabling frequency scaling of fade statistics to be modeled between UHF and Ka-band. In Alaska, measurements indicated that for a low elevation angle of 8°, a margin of 25 dB is required for 90% availability.

Dr. Vogel (left), University of Texas, and Dr. Goldhirsh (right), Johns Hopkins University, plan their central Maryland mobile propagation campaign for their mobile propagation van.

In another experiment, Georgia Tech characterized the impact of propagation impairments on candidate modulation schemes such as CDMA and phase-shift keying, which are being considered by future, commercial Ka-band systems [107]. JPL also measured and characterized the fading characteristic of the Ka-band channel in various suburban environments as they conducted land-mobile user application experiments [56].

Fade Compensation Techniques

Two primary methods for overcoming the severe rain fades at Ka-band frequencies are to increase the terminal-to-satellite link margins, or to provide earth station spatial diversity. Earth station diversity, although very effective,

169

is cost-prohibitive for low-data-rate user terminals because of the need for two terminals with an interconnecting link. To increase the link margin, most geostationary satellites find it effective to heavily code the links for error correction. For a regenerative satellite with base band switching, the uplink burst transmissions needs to be decoded (such as on ACTS), which consumes substantial onboard power.

Dr. Robert K. Crane, University of Oklahoma.

To reduce the amount of onboard power required for a concatenated code, adaptive fade compensation can be implemented in such a way that the inner code (normally a convolutional code) is not applied during non-fade periods. The system monitors the signal strength for each terminal and, during large rain fades, adaptively applies the convolutional coding. Since only a very small percentage of terminals in a multi-beam satellite will be encountering significant rain fade at any one time, significant power savings can be attained for the digital onboard demultiplexer, demodulator, and decoders (up to half of the power required) using this technique. This technique also provides much larger satellite channel capacity as discussed in Chapter 2, "Satellite Technology."

Adaptive Uplink and Downlink Fade Compensation ACTS tested the reliability of adaptively applying coding and burst rate reduction on the uplinks and downlinks for the VSAT network [43, 44]. The technique is described in Chapter 2, "Satellite Technology." ACTS was the first satellite system to use rain fade compensation on an operational basis. The data collected and analyzed by LeRC, using the extensive VSAT network operating with the base band processor, showed that the adaptive rain fade compensation system was very reliable. The protocol automatically detected fades and provided 10 dB of additional margin to maintain the BER at the service requirement. In no case did the system fail to successfully implement the fade compensation once the fade implementation threshold was crossed. The transitions to and from coded operations were accomplished without any loss of data.

Implementing adaptive rain compensation techniques on future communication satellites has the potential of significantly reducing the communication payload's weight and power. AstroLink, a Ka-band system proposed by Lockheed Martin, plans to use adaptive fade compensation on the uplink for just that reason. In a system such as AstroLink, the 1/2 rate convolutional code on the uplink is only applied to those transmissions that are being faded. Since a very small percentage of user terminals encounter rain fade at any one time, most of the uplink transmissions are made without the 1/2 rate coding. As a result, the user's uplink data rates are reduced by an approximate factor of two, as is the necessary onboard power for digitally demultiplexing, demodulating, and decoding the uplink channels.

Transmitter Power Control Adaptive downlink power control was considered for conserving power on board the ACTS satellite. Using this scheme, the downlink, 20 GHz TWTAs were to run in a back off condition, except when terminals in a spot beam encountered significant rain fade. When faded, the RF power for the affected beam/TWTA would be increased to compensate for the fade. After the ACTS TWT was developed, it was found that the DC power savings for a regenerative satellite using this method with base band switching were not great enough to justify the added complexity of implementing the method. Therefore it was never incorporated on ACTS. However, newer TWTAs that have greater efficiencies in the back off mode may make this method worthwhile.

Uplink power control is a technique that most satellites employ because of the need to maintain a relatively constant signal level on the satellite in the presence of various fading conditions along the propagation path. This is true whether the type of satellite is regenerative or bent pipe. COMSAT conducted an experiment to investigate the limitations of uplink power control in early 1995 [108]. The power control experiment involved estimating the uplink fade

171

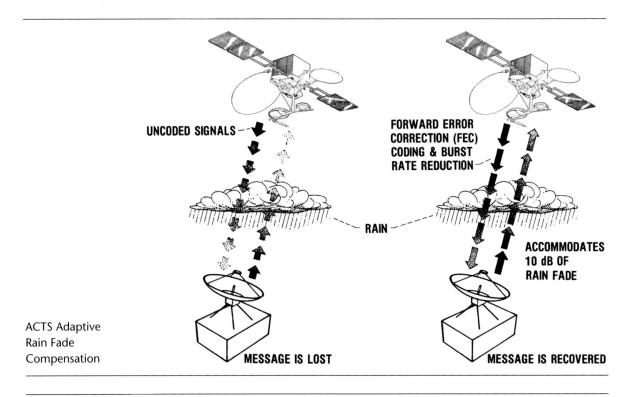

ACTS Adaptive
Rain Fade
Compensation

20 GHz rain fade
for T1-VSAT on
1/20/95 (four fade
compensation
events occurred at
7:39:45, 7:58:12,
8:20:08, and
8:57:43 GMT).
Each data point is
an average of 125
ms measurements
over a one-minute
interval. The
measurements are
made by the VSAT
in-channel
estimator.

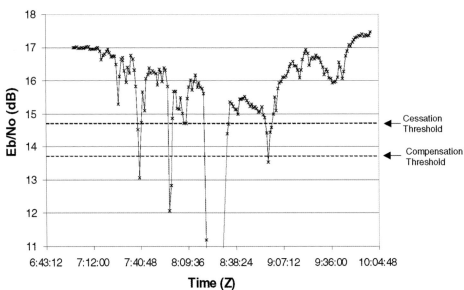

at an earth terminal by using a downlinked signal and increasing the transmitting power of the uplink carrier to compensate for the estimated 30 GHz uplink fade. The ACTS transmitted beacon signal in the uplink frequency bands was used to precisely measure the actual uplink fade. The experiment ran for six months. Using this fade detection method, the uplink power control accuracy could only be maintained within approximately +/-2.5 dB over a range of fades up to 18 dB at 30 GHz. The error increased as the fade depth increased. The main reason for the error is that the frequency scaling models were not able to accurately predict the 30 GHz uplink fade from the measured fade at the 20 GHz downlink frequency. The control algorithm also introduced errors.

Rather than use the open-loop power control scheme described in the previous paragraph, the proposed commercial regenerative satellites with base band switching actually plan to monitor the signal for each TDMA burst that arrives at the satellite using the onboard demodulated output. The estimated accuracy for measuring the uplink fade via the onboard processor is +/-1 dB, which is a considerable improvement over the method discussed in the previous paragraph.

Site Diversity Wide area site diversity entails the use of separated earth terminal sites, all connected into the same network, to combat rain fade. Rain cells are generally less than 10-20 km in length and oriented in certain directions. Site separation must be such that the same rain cell will not significantly rain upon two or more earth terminal sites. Under these conditions, the fading at one site tends to not be correlated with the fade at other sites. Terrestrial interconnections among the sites allow redistribution of traffic assigned to the faded earth terminal to those not currently undergoing fading.

Site diversity has a significant advantage in improving link availability in the presence of severe rain. COMSAT and Johns Hopkins University conducted a site diversity experiment using three sites more than 30 kilometers apart in Maryland and Virginia [109]. The key objectives of this experiment [110] were:

- to develop a diversity site-switching methodology based on monitoring propagation conditions at each terminal

- to investigate network control aspects required in diverting traffic between terminals, to minimize call blocking or packet delay

Testing over three months demonstrated the viability of wide area diversity as a means of combating fades and increasing the availability of VSAT terminals. The average time to complete a switchover was three seconds. The range of variation on switching time was very small, and the maximum value observed was five seconds. Data buffering had to be provided to avoid data loss during the switchover.

173

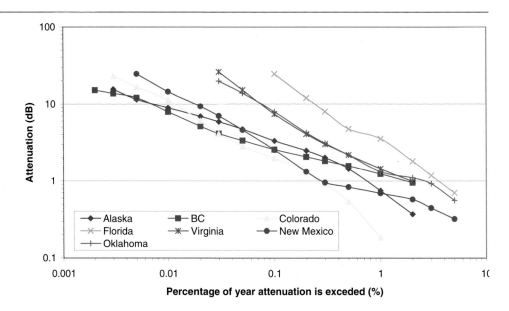

Five-year ACTS average annual rain attenuation at 20.2 GHz for seven sites. The measurement period is for 1994 – 1998 [105]. Attenuation is with respect to clear sky corrected for wet antenna effect.

Data provided courtesy of Dr. Robert K. Crane of the University of Oklahoma.

Water on the Antenna

Wetting of antenna surfaces by rain or snow causes additional signal losses [63, 111, 112]. Wetting consists of a thin water coating and beads of water on the VSAT's antenna reflector surface and/or on the antenna feed horn window. The amount of water coating on a surface depends upon the material, surface roughness, elevation angle, and exposure of the surface to aging. Water on the antenna surface, besides attenuating the signal, causes scattering losses due to the raindrops and creates a distorted reflector surface that reduces the antenna gain. The water coating on the feed window distorts the electric field distribution of the feed horn, causing an attenuation of the signal traversing the window.

Reflection from the ACTS propagation terminal antenna surface was complicated by the existence of a crinkled plastic dielectric coating over the conducting surface of the reflector. This dielectric coating acts as a spacer separating the partially reflecting water layer and droplets from the conducting surface. Because the plastic surface was not smooth, the antenna collected a small amount of water on the uneven surface. The ACTS propagation feed window was coated with a hydrophobic coating, however, so its surface shed water very readily.

The problem of water on the surface of the antenna was considered during the Olympus propagation experiments. None of the antennas used in the Olympus experiments produced attenuation values as high as those observed in ACTS. The Olympus antennas were of a different design and did not have a crinkled plastic surface covering the reflector surface.

Various tests were conducted on the ACTS propagation terminals by hand-spraying water on the surfaces. Based on those tests, it was estimated that during periods of heavy rain the wet antenna could produce as much as 3 dB of attenuation at 20.2 GHz and 5 dB at 27.5 GHz, in addition to the path attenuation. During periods of condensation or dew on the antenna, attenuation values as high as 4 dB have been observed at 27.5 GHz.

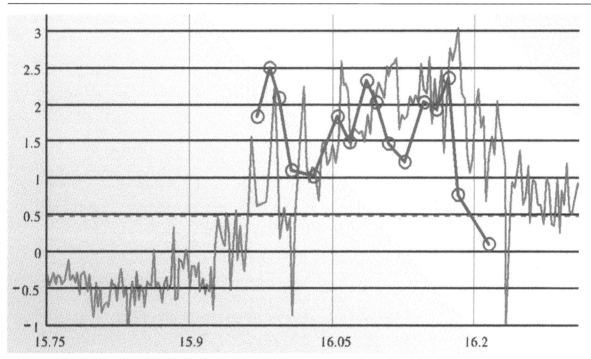

Wet antenna effect: Two identical antennas during rain event, with one shielded from the rain. Both antennas are measuring the same beacon signal under moderate rain conditions. Red curve is wet minus dry antenna attenuation with the vertical axis being differential attenuation in dB. Blue curve is rain rate with the rate being ten times the vertical axis in mm/hr. The horizontal axis is time in hours.

Roberto Acosta [112] performed comparison measurements using two ACTS antennas—one shielded from the rain and the other not shielded. His fairly extensive measurements at 20 GHz show that the:

- maximum antenna wetting factor was between 3 and 4 dB, and

- higher antenna wetting attenuation tended to occur at rain rates of 10–40 mm/hr.

175

It should be noted that at rain rates above 40 mm/hr the measurements might have been invalid due to the dynamic range limitation of the measurement equipment. The magnitude of the effect of water on the antenna reflector and feed surfaces is a function of the antenna design. A better design should reduce the attenuation observed and measured in these ACTS tests.

Depolarization

An understanding of the depolarizing characteristics of the earth's atmosphere is particularly important in the design of frequency reuse communication systems employing dual, independent, orthogonal-polarized channels in the same frequency band to increase capacity. Frequency reuse techniques—which employ either linear or circular polarized transmissions—can be degraded by the atmosphere through a transfer of energy from one orthogonal-polarized state to the other, resulting in interference between the two channels.

Rain-induced depolarization is produced from a differential attenuation and phase shift caused by non-spherical raindrops. As the size of rain drops increases, their shape tends to change from spherical (the preferred shape because of surface tension forces) to oblate spheroids with an increasingly pronounced flat or concave base produced from aerodynamic forces acting upward on the drops.

A second source of depolarization on an Earth-space path—in addition to rain—is the presence of ice crystals in clouds at high altitudes. Ice crystal depolarization is caused primarily by differential phase shifts rather than differential attenuation, which is the major mechanism for raindrop depolarization. Ice crystal depolarization can occur with little or no co-polarized attenuation. The amplitude and phase of the cross-polarized component can exhibit abrupt changes with large excursions.

During 1999 and 2000, atmospheric depolarization effects were measured using the ACTS beacons. The contribution of ice depolarization to the total depolarization on a radiowave link is difficult to determine from direct measurement, but can be inferred from observation of the co-polarized attenuation during depolarization events. As shown by the next figure, it can be assumed that the depolarization which occurs when the co-polarized attenuation is low, (i.e., less than 1 to 1.5 dB), is caused by ice particles alone, while the depolarization which occurs when co-polarized attenuation is higher can be attributed to both rain and ice particles.

For satellite links using signals at both polarizations and the same frequency, the effect of depolarization can be significant. Although most systems are designed to have polarization isolation of 27–30 dB, depolarization events due to the atmosphere can readily reduce that isolation effect by 8–10 dB.

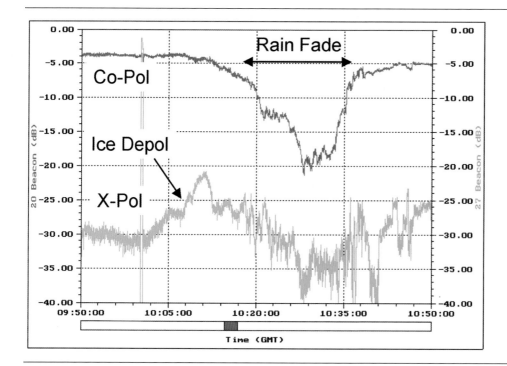

Ice depolarization event recorded using a terminal located in Virginia that was measuring the ACTS beacon signal at 20.185 GHz. Measurements made by Cynthia Grinder and Glenn Feldhake of ITT in cooperation with Roberto Acosta of the NASA Glenn Research Center.

Modeling

Extensive experimental research has been performed on the direct measurement of propagation effects on Earth-space links since the late 1960s. In parallel with this measurement activity, various propagation models have been developed for the prediction of RF attenuation. Virtually all of the propagation models use surface measured rain rate as the statistical variable and assume an aR^b type approximation between rain rate R (millimeters per hour) and rain-induced attenuation along a unit path length between the user terminal and the satellite. a and b are frequency- and temperature-dependent constants. Prediction of the percentage of time exceeded versus attenuation for a set period (one year or month) is a two-part process. First, a model of the rain rate versus percentage of time exceeded is needed for the geographical location of interest. Given the rain rate model, an integrated attenuation model over the path length must be used to predict the associated attenuation.

Many models for predicting rain attenuation are available [113]. A 1996 study of ten of these models by the International Telecommunication Union

177

(ITU) (using 186 station years of propagation data in the ITU database) found that no single model was best in consistency and accuracy.

One of the key activities of the ACTS propagation program was to compare the results of the measurements with these various prediction models. Robert Crane and Asoka Dissanayake compared the ACTS results [114, 115] with a combination of four attenuation-prediction models and three different, rain-rate prediction models. The Crane-Global rain climate model [116], combined with the Dissanayake, Allnutt, Daidara (DAH) attenuation-prediction model [117], gave the best results.

Using 21 station years of new ACTS data taken exclusively in the Ka-band across North America, Glenn Feldhake, Lynn Ailes-Sengers, and others, compared the ACTS measurements with the same ten models used in the ITU comparison [113] in 1997. In general, they found that the model errors were slightly greater than typical (30% to 40%). Based on the ACTS data, the ranking was not highly correlated with the previous ITU model comparison, with the exception that the DAH was best. It should be noted that the ITU has adopted the DAH model, therefore the ITU seems to agree with the Feldhake comparison.

ACTS Propagation workshop attendees at the El Segundo, California meeting in 1997.

The Crane-Global RF attenuation model [118] is based on the physics of attenuation by rain and predicted rain-rate probability distributions. Clouds, as well as antenna wetting effects, are not included. The above comparison shows that the Crane-Global model attenuation predictions were consistently low. Other models are more empirically derived from measurements and inherently include these other effects. In order for a physically based attenuation model to do a better job at prediction, a model for antenna-wetting effects needs to be included. Roberto Acosta developed an antenna-wetting effect model based on measurements [112, 119]. Using this model, Robert Crane [120] adjusted the ACTS rain attenuation measurements to remove the attenuation due to antenna wetting. Crane reported a much-improved agreement between his two-component model and the wet antenna, adjusted ACTS rain attenuation measurements. Since the effect of antenna wetting is significant, future user terminals need to incorporate features that minimize those effects.

Another important consideration will be the validation of the models at low elevation angles. Very little data exists below an elevation angle of 20°. Low elevation angle data is particularly important for tropospheric scattering. The ACTS site at the University of Alaska in Fairbanks has an elevation angle of 8° and will provide five years of measurements from which to validate low elevation angle models.

Because of the severe propagation effects at Ka-band, it is important to refine the old models and develop new models that improve the capability so the system designer can better predict needed link margins. The ability to realistically combine the effects of rain, gaseous and cloud attenuation, scintillation, and so forth, should also be included.

In the design of future commercial systems, every dB is critical. Insufficient margins will reduce availability, and excess margins will increase earth terminal size and cost.

Concluding Remarks

The work done by the ACTS propagation researchers in characterizing the Ka-band frequency and its susceptibility to rain and other atmospheric and transmission path impairments is critical for the development of the next generation of satellites that plan to use these frequencies. One of the key features of the ACTS propagation program was that all seven ACTS sites used the same collection hardware and employed the same software to process raw data. Thus, the data is easily compared and is equally valid for all the rain climates in which studies were conducted.

The ACTS Ka-band propagation measurement data set is the largest set of Ka-band measurements in North America. Empirical cumulative distribution for satellite-to-space path attenuation relative to clear sky values has been compiled for 35 path years of data. The five years of measurements provide a statistically adequate time period for data collection. The five years of data produce a distribution estimate with only an 11% uncertainty. To date, only three locations worldwide have produced observations of five years of more. The ACTS propagation data should more than triple the amount of data available.

These propagation statistics are for two frequencies—20.2 and 27.5 GHz—with elevation angles ranging from 8 to 52°, latitudes ranging from 28 to 65°, and five different rain climate zones. In addition, the accuracy in prior propagation measurement campaigns was generally less than 1 dB. ACTS measurement precision was about 0.5 dB, with a rms measurement error of 0.1 to 0.2 dB. For locations within CONUS, the dynamic range was better than 20 dB.

Fade compensation techniques will play a critical role in future commercial exploitation of the Ka-band frequencies. The research performed using ACTS on fade mitigation techniques has proved that adaptive coding can be effective in compensating for severe rain fades and saving spacecraft weight and power. The initial summary comparison of ACTS Ka-band data with prediction models does indicate a need to revise the models. The ACTS data will accelerate efforts to improve existing models for predicting atmospheric impairments of Ka-band satellite links. This will meet the urgent needs of the satellite communication industry as it plans for the commercial utilization of Ka-band frequencies.

CHAPTER 6

THE DEVELOPMENT OF ACTS

Most of the ACTS technology, described in Chapter 2, "Satellite Technology," had never before been put in space. Its development was a major technical challenge, filled with frustration, stress, and risk-taking, but offering major opportunities and rewards. It required a large government/private industry team integrating a variety of disciplines—including science, engineering, planning, finance, and human resources—to accomplish an important goal that made a real difference. In short, ACTS was being where the action is in the development and application of exciting new technologies and processes.

The keys to a successful program are no secret but bear reiteration here to point out the complexities of the job, and serve as a basis for understanding the benefits and the mistakes. Some of the main components of a successful program, according to A. Thomas Young (former president of Lockheed Martin and center director of Goddard Space Flight Center [121]), are:

- finding good, experienced people

- instilling attention to detail

- building adequate reserves

- avoiding political consideration

- putting quality first

Good, experienced people Getting the right people on the team is what separates successful projects from those that fail. A good engineer is one who has designed something, had it built, and seen it fail. There have to be enough experienced personnel (managers, engineers, technicians, planners, etc.) on the team who will make certain that the same type of mistakes made on previous projects won't be repeated. For those who don't have the experience, the proper training must be provided.

Attention to detail Space projects rarely fail because of large flaws. It is usually the overlooking of seemingly small details that dooms otherwise sound programs.

Adequate reserves Repair shops are few and far between the earth's surface and geostationary orbit. Working on an experimental program such as ACTS requires adequate margins in funding, scheduling, spacecraft performance, and so on.

Political consideration While every project begins with technical need, political considerations soon tempt projects to weave in other factors that will raise cost and risk. To avoid such influences, the project needs an inviolate shield around it.

Quality first Performance tests have to be built into each step of the project. Quality cannot be balanced with other variables—it must be the top priority.

In this chapter we will review the development of ACTS and provide a scorecard on how well the financial, schedule, and technical objectives were accomplished, as well as the reasons for the successes and failures. Using this process, we will see how well these key steps were implemented to achieve a successful project.

One Strike For (and Three Strikes Against) A Successful Development

Motivation is another key factor that makes a big difference in whether or not an endeavor is successful. Norman R. Augustine, former chief executive officer of Lockheed Martin nicely summed up the role that motivation plays in his book entitled, *Augustine's Laws*. [122] He stated that "Motivation makes the difference. In sports this is sometimes equated with 'mental toughness'—how else can one explain the numerous occasions each season when a team soundly beats another only to find itself thrashed by the loser a few weeks later? Motivation will almost always beat mere talent."

At the outset, there were major strikes against ACTS that were significant negative motivation factors:

- First, the federal administration never supported the program and the Office of Management and Budget zeroed out the budget for ACTS before the development contract was awarded and during the following four years. These actions continuously put a negative cast over the project.

- Second, there were many people within NASA who opposed ACTS because it was not a NASA mission (i.e., it provided technology for commercial purposes only). As a result, many people within NASA considered it a low-priority project.

- Finally, there was no planned follow-on to ACTS that could be used by NASA LeRC as a carrot for getting good contractor performance. When allocating personnel resources to jobs, contractors frequently assign the best people to those jobs that will bring in repeat business at a handsome profit. For ACTS, assignment of experienced contractor personnel was a significant problem.

The major spacecraft contractors (Hughes, Ford Aerospace (Loral), RCA AstroSpace/ Lockheed Martin) competed aggressively with each other for both commercial and government contracts. Hughes opposed the development of ACTS and lobbied against the development contract award to the RCA contractor team that consisted of first tier subcontractors, TRW and COMSAT. This action by Hughes was actually a positive motivational factor for RCA. It served to increase the resolve of Charles Schmidt (the head of RCA AstroSpace at the time) to make ACTS successful. In the end, it was largely due to actions by Charles Schmidt that led to the successful development of the ACTS program. He placed a high value on the technology and went to bat for the program both internally and externally.

The Initial ACTS Development Contract

The NASA Lewis Research Center (LeRC) awarded the prime contract for the spacecraft, the master control station, and two years of in-orbit operations to RCA (which in time merged successively with GE, Martin Marietta, and Lockheed Martin) on August 10, 1984, for $260 M. The contractor team consisted of RCA as the system integrator and provider of the spacecraft bus, TRW as subcontractor for the communications payload, and COMSAT as subcontractor for the NASA ground station (NGS) and master control station (MCS). The $260 M did not include the cost of the shuttle launch or the booster to place the spacecraft in a transfer orbit from low-earth, shuttle altitude to geosynchronous altitude. The cost for the launch and the PAM-A booster was not included in the prime contract.

At the time of the contract award, the LeRC cost estimate for development of the ACTS system was $329 M. This estimated cost included the prime contract, the LeRC contingency (approximately 15%), and other development costs including support for long-lead technology development. This $329 M total did not include the additional program funds levied by NASA's headquarters for taxes that occur at the agency level.

The ACTS system design, development, and launch preparation period was initially planned for 60 months, with the launch of the spacecraft projected for September of 1989. The normal commercial communication spacecraft at the time was being developed in about three years compared to this five-year scenario. The extra time for ACTS was judged to be necessary to overcome new technology hurdles, to develop an extensive communications payload engineering model, and to conduct a three month, pre-launch systems test for checking out the complicated network between the onboard processor, the master control station, and the user terminals. The engineering model for the payload was included to remove development risk. The ground/space-

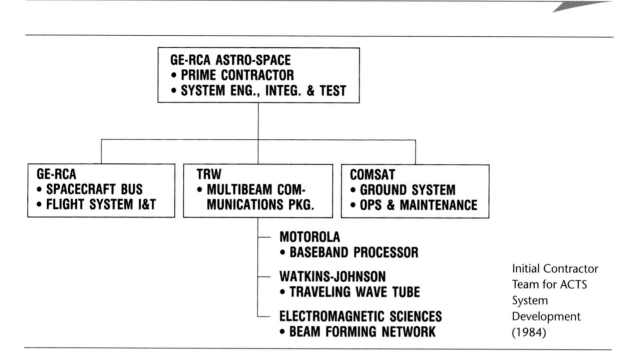

Initial Contractor Team for ACTS System Development (1984)

craft system test is not normally conducted when producing a commercial bent-pipe type spacecraft, but was deemed necessary for this first generation, onboard processing satellite.

Because of the high technical risk and one-of-a-kind nature of the ACTS program, the only practical procurement mechanism was a cost-plus-fee contract. Under this type of contract, the government takes on all cost risks. Because the technology was being developed for commercial application, the fee allotted for the contractors was fixed at 5.5%, which was considerably below the 12–13% normally collected. With this fee arrangement, the contractors were, in essence, making a direct contribution toward the funding of the program. That contribution, at the time of award of the contract, was approximately $20 M (the fee difference between the normal 13% and the ACTS contract 5.5% fee). In addition to the lower-than-normal fee, RCA contributed approximately $2 M for the steerable antenna that was added to the spacecraft after contract award. This steerable beam antenna proved to be a valuable addition and was extensively used during ACTS operations. All the major contractors on the team contributed their own funds to purchase terminals and/or to support the planning, solicitation, and conduct of experiments or user trials. This amounted to approximately $4–5 M more in contractor contributions.

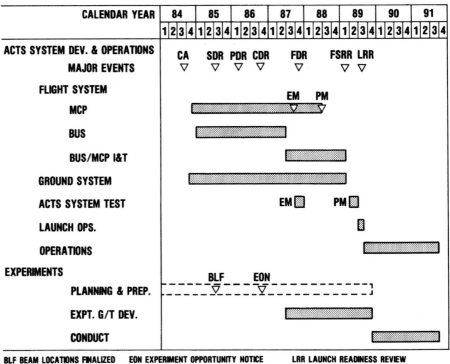

CALENDAR YEAR	84	85	86	87	88	89	90	91
	1 2 3 4	1 2 3 4	1 2 3 4	1 2 3 4	1 2 3 4	1 2 3 4	1 2 3 4	1 2 3 4

ACTS SYSTEM DEV. & OPERATIONS
MAJOR EVENTS — CA, SDR, PDR, CDR, FDR, FSRR, LRR

FLIGHT SYSTEM
MCP — EM, PM
BUS
BUS/MCP I&T

GROUND SYSTEM

ACTS SYSTEM TEST — EM, PM
LAUNCH OPS.
OPERATIONS

EXPERIMENTS
PLANNING & PREP. — BLF, EON
EXPT. G/T DEV.
CONDUCT

BLF BEAM LOCATIONS FINALIZED	EON EXPERIMENT OPPORTUNITY NOTICE	LRR LAUNCH READINESS REVIEW
CA CONTRACT AWARD	FDR FINAL DESIGN REVIEW	MCP MULTIBEAM COMMUNICATIONS PACKAGE
CDR CRITICAL DESIGN REVIEW	FSRR FLIGHT SYSTEM READINESS REVIEW	PDR PRELIMINARY DESIGN REVIEW
EM ENG. MODEL	I&T INTEGRATION & TEST	PM PROTOFLIGHT MODEL
		SDR SYSTEM DESIGN REVIEW

Initial ACTS
Development
Schedule

The ACTS Development Scorecard

At the bottom-line level, the technical, schedule, and cost performance is readily expressed:

- The launch occurred on September 12, 1993—some four years after the original planned date

- NASA LeRC's cost for the program grew from $329 M to $481 M

- All technical objectives were met or surpassed

- The satellite's lifetime expectancy was exceeded by more than three years. Only minor spacecraft anomalies occurred, none of which decreased its communications capacity

A quality product was produced at considerably more money and time than initially anticipated. Development of high-risk technology is seldom produced in less time or at a lower cost than planned. The remainder of this chapter will discuss the primary reasons for cost growth, schedule extension, and the attainment of a quality product.

Elements of Cost Growth

Technical Changes

From the start, the LeRC project management philosophy was to write the ACTS technical requirements in terms of general functional capability, without dictating the detailed technical approach or changing the technical requirements. This latter philosophical item meant resisting outside influence to add goodies beyond the original scope. By taking this approach, the NASA LeRC project management team hoped to ensure its integrity (by not trying to get added features for no or little increase in cost). This objective was almost completely accomplished. Over the lifetime of the program, the technical scope changes to the development contract resulted in a net reduction in cost.

Early in the development program, the contractors suggested (and NASA implemented) technical changes to reduce program costs. These included deleting a full-up engineering model for the multi-beam antenna, a spare flight model BBP, and a diversity earth station. In addition, the base band processor's downlink burst rate and total throughput were cut in half and the number of spot beam locations was significantly reduced. In total, more than $25 M in technical scope was removed from ACTS.

Funding Constraints

The control of the amount of funding any program receives each year is in the hands of the NASA administrator, the federal government administration, and the U.S. Congress. The administration's Office of Management and Budget (OMB) zeroed-out funds for ACTS in its yearly budget requests to Congress in 1985 and 1987-1990. Congress was then put in the position of reinstating funding for the program over the objection of the administration. This process frequently resulted in less money being appropriated than originally planned, with a resultant slip in the development schedule and an overall increase in total project costs. The reason that total project costs grew with a stretch-out in funding is that many people (such as project controllers and system engineers) were required to remain with the project as long as it was in the development stage. Frequent reassignment of key individuals to other projects, with the loss of their knowledge and ownership, was not practical. With the large number of subcontractors, the effort to create a new plan after each funding cut also took considerable time and effort (money).

The first time that funding was zeroed-out was in December of 1983, just prior to the planned contract award. As explained in Chapter 1, "Program Formulation," the administration chose to do this when Hughes submitted their filing for their own Ka-band satellite and claimed that the NASA program was unnecessary (Hughes never followed through and actually built the spacecraft). This OMB action caused the development contract award to be delayed 8 months—from December of 1983 until August of 1984. When the funds were finally reinstated for the project, they were done so at a lower level than the development plan required. Worse than the effect of the reduction in funding was probably the impact on contractor personnel. Many of the people who had worked on the proposal were no longer available for assignment to the project eight months later. Therefore, new individuals who were not familiar with the project and who could not be held responsible for the proposed development plan had to be assigned. This is not the way to start a project!

Fiscal year funding constraint were imposed upon the development program four times, causing schedule stretch and increases to the total program cost. The total project cost growth due to program stretch-out was $53 M.

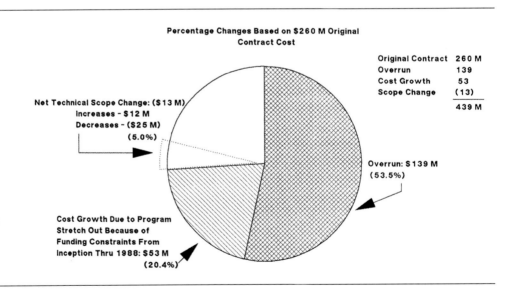

Percentage Changes Based on $260 M Original Contract Cost

Original Contract 260 M
Overrun 139
Cost Growth 53
Scope Change (13)
———
439 M

Net Technical Scope Change: ($13 M)
Increases - $12 M
Decreases - ($25 M)
(5.0%)

Overrun: $139 M
(53.5%)

Cost Growth Due to Program Stretch Out Because of Funding Constraints From Inception Thru 1988: $53 M
(20.4%)

Elements of ACTS Contractor Cost Growth as of January 1988

Overrun

Contractor overrun was $139 M. The overrun, which occurred mostly during the second and third years (1986 and 1987) of the development contract, was due primarily to the TRW communication payload subcontract. LeRC project management believed that one of the main reasons for TRW cost growth was that TRW and their subcontractors underestimated the true development cost in their proposal. Such underestimating has always plagued the government in cost-plus programs, and it appeared that ACTS was no exception. This was not totally the fault of the contractor. The contractors knew the government's target price for ACTS and made sure they met it in their proposals.

Another reason for the TRW overrun is believed to be the low priority TRW placed upon the program. At the time of the ACTS contract, TRW had many classified programs that provided them with long-term revenue and profits. ACTS, being a one-of-a-kind spacecraft, appeared to have significantly lower priority than those built for military programs. Few seasoned veteran engineers and managers were assigned by TRW to the ACTS project. Having more veteran engineers on the program would have gone a long way toward eliminating many technical and management problems, with a resultant reduction in overrun.

The Capped Program

As the overrun grew, Congressional support for the program started to wane. By late 1987, it was clear that if the program was to continue, some sure way to limit cost growth was needed. The choice that satisfied Richard Mallow, the chief congressional staffer on the House appropriation committee in charge of NASA's budget, was to cap the total NASA cost for the project. This cap was to apply to all levels, including NASA headquarters, NASA LeRC's project group, and each of the development contractors. All contracts were to be changed from cost-plus into ones that were capped. Mallow desired that the ACTS program be capped at $499 M.

With three years of work completed, a lot of the risk associated with the ACTS development was removed or better understood. As a result, the LeRC project management felt that conversion to capped contracts was feasible. At this time, TRW had already incurred costs for approximately $153 M, and wanted $72 M more to complete the payload under a capped contract—which they were very hesitant to accept (the original contract value for the communication payload was only $106 M).

When all of the costs were received from the contractors for completing the project under capped contracts, the total exceeded the $499 M Mallow target. For this reason it became necessary to restructure the development team to realize additional savings. The program shifted from a prime contract to two associated contractors.

First, it was decided to drop TRW as the subcontractor for the communications payload and have GE AstroSpace complete the payload using TRW's second tier subcontractors (GE projected they could finish the payload for $14 M less than TRW). The second part of the restructuring was to have NASA LeRC

189

become the integrating contractor, with one contract with GE for the spacecraft and another contract with COMSAT for the ground segment. This eliminated considerable GE costs for managing the integration of the spacecraft and ground segment activities. Net savings were realized because NASA LeRC was able to perform the integration tasks without increasing the program staff.

With this shifting of responsibilities, the completion of ACTS development and two years of in-orbit operations became possible within Mallow's target of $499 M. After subtraction of the NASA headquarters' taxes, the total amount of funds available to the project office at NASA LeRC for ACTS development and the first two years of operations was $478 M. This amount, however, allowed NASA LeRC to hold only several million dollars for contingency purposes.

Much of the credit for saving the program must go to Charles Schmidt and the original RCA AstroSpace team members who were motivated to honor their original commitments and take on considerable risks to complete the payload development under a capped contract. The acceptance of capped contracts by GE, COMSAT, and Motorola showed their belief in the value of the ACTS technology. The conversion worked. All parties reached agreement in January 1988. In a letter from Senator William Proxmire (D-Wisconsin) and Representative Edward P. Boland (D-Massachusetts) (chairman respectively of the Senate and House Subcommittees on HUD-Independent Agencies) to James Fletcher, the administrator of NASA, Congress agreed to the $499 M capand provided $76 M for both FY 1989 and 1990. From that time forward, cost growth concerns lessened and everyone turned to concentrate on the task of developing a quality product.

AstroSpace team at System CDR, January 1990: (left to right) John Graebner, Frank Boyer—2nd Program Manager for AstroSpace, Richard Moat—Motorola, William Cashman, Josh Shefer, George Beck, Frank Gargione, Al Little, J. Dearden, J. Dyster and Paul Levine

The capping of the program provided some measure of assurance to the development team that their funding would stabilize. One year later (in January 1989) however, OMB once again zeroed-out the ACTS funding in the administration's FY 1990 budget request to Congress. Once again the Congress reinstated the funds for ACTS. This was OMB's final attempt to cancel the ACTS program.

Complexities Associated with a Large Program

When the development contract was awarded in August of 1984, the contractor team consisted of RCA as the prime, TRW and COMSATas first-tier subcontractors for the spacecraft communications payload and the ground segment, respectively. Below the first-tier subcontractors, there were approximately 15 major second-tier subcontractors. Some of the second-tier subcontractors—such as Motorola for the base band processor ($46 M), and EMS Technologies for the beam-forming network ($9 M)—were developing high-risk, advanced technology components.

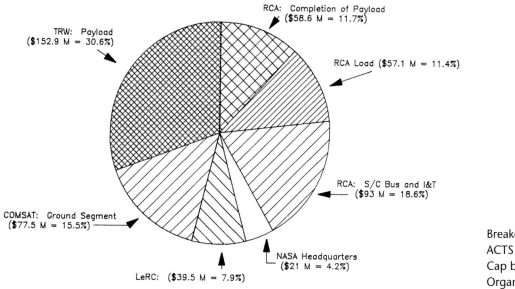

Breakdown of ACTS $499M Cost Cap by Organization

For a first generation system like ACTS, layers of contractors can create major complications. For instance, TRW was awarded a definitive contract by RCA on February 25, 1985. EMS Technologies (a subcontractor for TRW) completed no substantive work on the beam-forming network until after it signed a letter contract with TRW on April 3, 1985. In August 1985, the ACTS system design review identified that EMS and TRW could not complete the multi-beam antenna system, including the BFN, for the allocated budget. To reduce costs, NASA's LeRC decreased the BFN scope as recommended by the contractors. The redefined contract between TRW and EMS wasn't signed until July of 1986, which was almost two years after the prime contract was awarded.

As the above example illustrates, the process of finalizing the requirements and contracts for a new system can be very complex and time-consuming. The elimination of RCA as the prime contractor and TRW as the payload subcontractor went a long way to decrease the high degree of organizational layering.

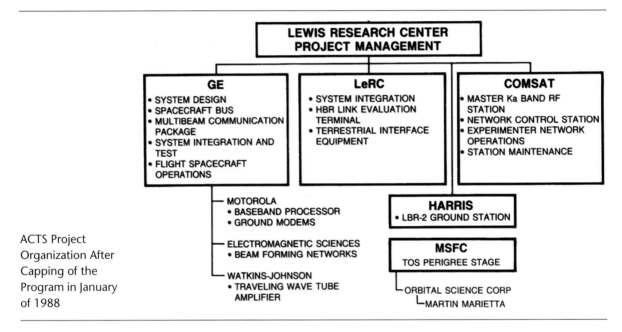

ACTS Project Organization After Capping of the Program in January of 1988

The restructuring of the development team in 1988 was certainly a correct step for improved performance. There had been considerable friction between the prime contractor (RCA), first-tier subcontractors COMSAT and TRW, and some of the second-tier subcontractors, which made problems more difficult to resolve and contributed to schedule delays. A committee consisting of executives above the program manager level from each contractor and NASA was established to resolve top-level issues and reduce friction. Even though it helped, it was not enough. When NASA took over the system integration role and, in effect, became the prime contractor, much of this friction was eliminated. In 1988, NASA engineers took over the lead of the working groups (RF, command and telemetry, and on-demand network control) that were responsible for coordinating the interface requirements between the spacecraft and ground system. NASA promoted as rapid a technical resolution of problems as possible, and then funded any resulting contractual scope changes using its contingency funds. Use of NASA contingency funds for this purpose went a long way toward improving relations between the financially pressed contractors.

The approach that LeRC took for managing the ACTS project was to have a strong technical, cost, and schedule team that could delve closely into all aspects of the contractor activities. The ACTS project organization mirrored that of the contractor team, and monthly performance meetings were held at all major contractor facilities including those of the principal, second-tier subcontractors. In addition, NASA project personnel participated in the standard design reviews including those for all first- and second-tier subcontractors. The idea was to have enough in-house knowledge and expertise to perform in-depth assessment of all issues, so that the government team could make independent but practical decisions, be proactive in solving problems, and help ensure a quality product. Initially the government project team did not have all the necessary skills needed to accomplish this, but it did by the time of the restructuring and that allowed it to assume the role of integrating contractor for the ACTS system. This role involved more than just managing the spacecraft and ground segment integration. It also included development for more than seven types of user terminals.

NASA LeRC team at System CDR in Jananuary, 1990: (left to right) Karl Reader, Rod Knight, Tom St. Onge, Richard Gedney—LeRC project manager, Howard Jackson, Pete Vrotsos, Dave Wright, and Ernie Spisz.

For each of the four times that funding constraints were imposed on the program, a six-month process took place where funding cuts were allocated at each contractor level, new longer schedules and increased contract values were negotiated, and a new launch date was established. This whole process distracted the development team from their main tasks and consumed a large number of man-hours. It also gave the contractors an opportunity (it is the government's management job to make sure it doesn't happen) to get well

193

regarding some of their own problems. One of the practices for good project management is to never take an action (such as funding cuts) that gives the contractor a reason for failing to meet development schedules.

The impact of funding constraints on a complex development program like ACTS can be disastrous. As bad as those impacts can be (on ACTS they increased the total cost by $54 M), top-level government administrators and the U.S. Congress seemed to ignore the consequences and just considered funding cuts an inevitable part of generating a yearly budget.

Completion of the Communication Payload

At the time that the program was capped in January of 1988, the flight base band processor was 85% completed by Motorola, the flight beam-forming networks were due to be delivered by EMS Technologies two months later, and the 20 GHz flight traveling wave tubes were completed and in test at Watkins-Johnson. In addition, 70% of the TRW flight drawings were released and the engineering unit models had completed test. The biggest deficiency was with the mechanical design of the multi-beam antenna (MBA), for which TRW had only completed 35% of the drawings.

The GE approach for completing the payload was to use the completed portions of the TRW design, making only those changes necessary to accommodate the GE manufacturing processes and test procedures or to correct design deficiencies. In addition, GE decided to use all the parts and material TRW had received. As far as completing the design of the MBA, GE felt that it was well within the capabilities of its engineers. Using this approach, GE forecast that they could complete the payload for almost $14 M less than TRW ($58 versus $72 M).

This was how the communication payload was completed. It is TRW's initial design that was completed and manufactured by GE. To TRW's credit, they completely cooperated in the transition of all design documents and material to GE. There were many skeptics who thought GE's plan would not be successful. But Charles Schmidt, head of GE AstroSpace, had confidence that his people could pull it off!

Major Increases in Scope

At the start of ACTS development in August of 1984, the spacecraft weight at beginning of life (BOL) in-orbit was estimated to be 2,367 pounds, with 730 pounds of that total being the communication payload. A margin of 347 pounds was available for weight growth as the design matured. In addition, only enough stationkeeping fuel for two years of operations was included. With this spacecraft weight, a payload assist module (PAM-A) was to be used to place the ACTS in the proper transfer orbit from shuttle altitude to geostationary altitude.

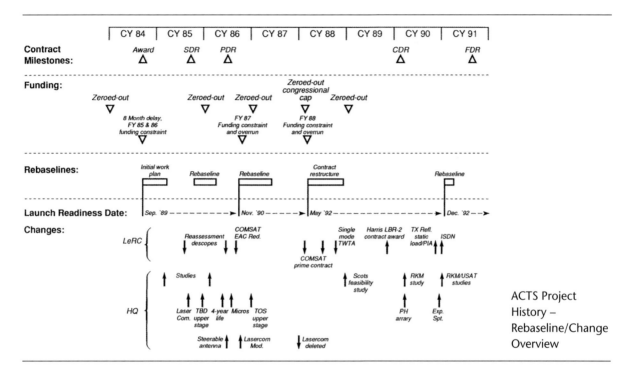

ACTS Project History – Rebaseline/Change Overview

At launch, the final spacecraft BOL weight was approximately 3,250 pounds, with five years of stationkeeping fuel and 116 pounds of excess margin. Of this total, the communication payload weight was 1,295 pounds, including a 33-pound steerable beam antenna that was added in 1986. The big increase in payload weight from 1984 to 1993 was largely due to changing the antenna reflectors' stowed position during launch. The original design had the reflectors folded across the top of the spacecraft to minimize its length. This was changed to include a major truss structure on top of the spacecraft to hold the reflectors and beam-forming networks during launch.

To accommodate the 3,250-pound weight, the Orbital Sciences' Transfer Orbit Stage (TOS) was used in place of the PAM-A. The TOS stage capability was more than required and it did not carry a full, solid propellant load. The switch to the TOS upper stage was made in 1985, with addition of the laser communications package (Lasercom). This package (which was being developed by MIT Lincoln Laboratories for the Air Force), definitely placed the ACTS weight outside the capability of the PAM-A booster. As it turned out, after the addition of the Lasercom and the switch to the TOS was contractually completed, the Lasercom funding was cancelled by the Air Force and the package was deleted in 1987. The combined Lasercom package and TOS change was the largest single scope increase to the contract, amounting to almost $9 M. By holding the scope increases to a minimum, NASA helped to contain the project cost and risk.

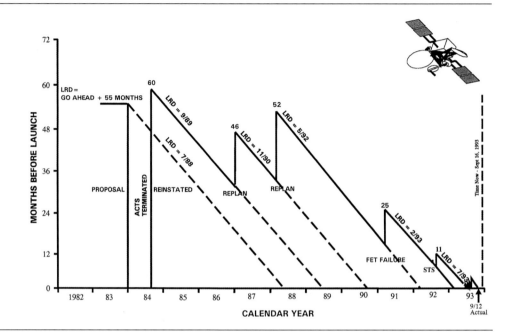

History of Launch Readiness Date (courtesy of Rod Knight, NASA)

Elements of ACTS Schedule Slips

The ACTS contract award was originally planned for December of 1983 with a launch in July of 1988. The actual launch occurred more than five years later on September 12, 1993. The combined schedule impacts due to program approval delays, funding constraints, cost overruns, technical difficulties, and launch vehicle availability were gigantic. The effects of some of these factors will be discussed in more detail in this section.

Funding Constraints and Overruns

A non-advocate review, headed by Dave Pine of NASA headquarters, evaluated the ACTS program plan in July of 1982 and concluded that a peak year funding of approximately $140 M was required to complete the 55-month development. When ACTS development was finally started in August of 1984, peak year funding was planned to occur for two subsequent years at approximately $110 M each year, with a launch 60 months later. The maximum funds allocated to ACTS in any one year were $85 M. With the yearly funding for ACTS always being less than required, and cost overruns increasing the total amount of funds required, it was impossible to meet the original 60-month development schedule.

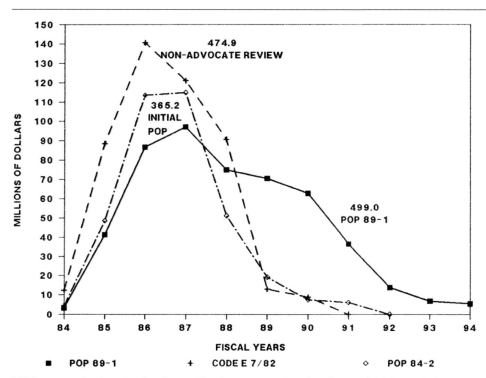

ACTS cost projections by fiscal year. Fiscal year starts October first and POP stands for Program Operating Plan. The non-advocate review estimates were made in July 1982 by a group of independent financial and engineering experts. This group estimated a total program cost of $474.9 M. The initial POP estimates were made by the ACTS project staff at LeRC in the second half of 1984 and predicted a total cost of $365.2 M. The 89-1 POP estimates were made by the ACTS project staff at LeRC in the first half of 1989 and predicted a total cost of $499 M. This plan which "capped" the total cost using fixed price contracts was successfully achieved. The costs in the last two years are for operations. Note that the non-advocate review estimates were close to the final achieved costs.

In 1986 and 1987 the launch date was slipped 14 and 18 months, respectively. Part of the slip that occurred at the end of 1987 was due to the program capping process that took considerable time to accomplish. Therefore, the funding constraints and the cost overruns were responsible for 32 months of slippage (projected launch date was moved to May of 1992).

Technical Difficulties

When the contractor team was restructured in January of 1988, schedule margins were put in place to ensure meeting the new launch date of May 1992. As it turned out, the margins were not large enough to accommodate several technical problems and a procurement difficulty.

One of the most unexpected impacts was related to the ordering of parts by Astro Space for both the spacecraft bus and the communication payload. For a spacecraft, part procurements represented a major activity. First, NASA required (for quality reasons) that all electronic parts that had not been flown before be procured under Class S equivalent standards. For each new part, the procedure required the generation of a requirements document that specified a high degree of inspection and testing by the manufacturer to ensure that the part would be reliable for use in space. The requirements document for each new part had to be approved by NASA, which added time to the procurement process.

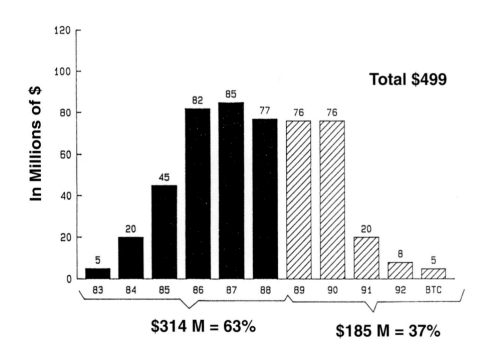

ACTS Appropriations by Fiscal Year (January 1988)

Second, the incorporation of the GE parts ordering system in place at RCA Astro Space (merger of the RCA and GE aerospace units was initiated in 1986) was still taking place and drastically slowed the process. Third, the parts ordering process for many components was not initiated on time due to management mistakes. Even though extra people were assigned to speed up the process, the end result was that the assembly of part kits for the manufacturing of electronic boxes and sub-assemblies was significantly delayed. It took months to resolve this problem, which completely eroded schedule contingency.

A second major schedule hit was due to a problem with one critical electronic part. In any spacecraft program, one of the worst things that can happen is finding a defective part after it has been installed in boxes that have completed the lengthy manufacturing/test cycle. This disaster happened to ACTS!

Throughout many electronic boxes in the communication payload, a field effect transistor (FET) was used in many hermetically-sealed amplifier modules. The laser-welded hermetic seal prevented unwanted atmospheric gas from entering the modules and causing damage by corrosion. After the many boxes that contain this FET had completed their manufacture/test cycle, it was discovered that the FET's expected life fell well short of the ACTS requirement. Resolving this problem meant procuring new FETs, disassembling the affected boxes, breaking the laser-welded seal, installing the new FETs, and reassembling and retesting the boxes.

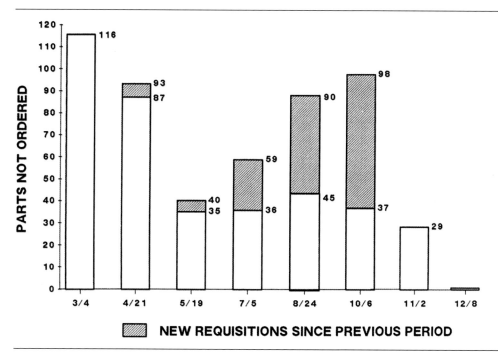

Tracking EEE Part-Ordering Status - Spacecraft Bus ("Parts Not Ordered" refers to part types, not total number of parts. Most bus EEE parts were heritage items that did not require special NASA approval to procure. Despite not having to obtain NASA's approval, it still took Astro Space more than one year to order the bus parts.)

Delays with part ordering and the lengthy FET refurbishment, coupled with manufacturing problems encountered with the multi-beam antenna were the major contributors to an additional nine-month slip in the launch date. This slip, which occurred in early 1991, extended the launch from May of 1992 to February of 1993 and was the fourth one to occur.

The final launch schedule slips were related to the unavailability of the shuttle. In September of 1992, NASA decided to expand the Atlantis shuttle refurbishment activities at Palmdale, California to include the incorporation of a docking capability with the Russian MIR space station. Although ACTS was manifested on Discovery, this caused a ripple effect in the entire shuttle manifest including the rescheduling of the ACTS launch from May of 1993 to July of 1993.

The spacecraft was delivered to NASA's Kennedy Space Center in February of 1993 to be mated with the TOS upper stage and begin final preparations for launch aboard the shuttle. The first launch attempt was made on July 17, 1993. It was scrubbed less than an hour before the planned liftoff, due to a faulty transistor in the launch system that armed a set of explosive bolts prematurely. On July 24, a second attempt was made. It was aborted 19 seconds before launch when a turbine in a shuttle auxiliary power unit (APU) didn't come up to speed properly. All further launch attempts were delayed until the APU was changed and the Perseid meteor showers had passed. During the August 12th (third) launch attempt, a sensor failed to indicate that propellant was flowing and the shuttle's engines were automatically shut down three seconds before liftoff. On September 12, 1993, at 7:45 AM EDT, ACTS was finally launched aboard shuttle Discovery on mission STS-51. ACTS arrived at its permanent location in geostationary orbit at 100 degree west longitude on September 28, 1993.

NASA S/C launch team at JSC mission control in Houston, TX: (left to right) Joe Nieberding, Glenn Horvat, John Collins, Richard Gedney, Erv Edelman, and Richard Krawczyk.

A Quality Product

Poor quality certainly costs money. This was vividly illustrated by the FET episode described in the previous section. Had the FET manufacturer produced a quality product, it would have saved a lot of money and potentially another embarrassing launch schedule slip.

To successfully produce a spacecraft, experienced people are needed who have learned from previous failures, know how to build it right the first time, and know how important it is to train and mentor the inexperienced members of the team. It also takes a total team commitment to many small details. The relationship between the customer and the contractor team building the system needs to be complementary and not adversarial. With adversarial relations, decisions can be made that sacrifice quality. Just as important, adequate testing must be done to demonstrate that the design is functional and the spacecraft will work in space.

Part of the Lockheed Martin team that built a quality product.

The lack of attention to details and team commitment is illustrated by the Hubble Space Telescope failure, which launched a flawed telescope into space. According to the NASA failure report [123],

> The most unfortunate aspect of this HST optical system failure, however, is that the data revealing these errors were available from time to time in the fabrication process, but were not recognized and fully investigated at the time. Reviews were inadequate, both internally and externally, and the engineers and scientists who were qualified to analyze the test data did not do so in sufficient detail. Competitive, organizational, cost, and schedule pressures were all factors in limiting full exposure of all the test information to qualified reviewers.

Notice the lack of team commitment and the presence of adversarial relations in this finding.

With all the controversy surrounding ACTS (the yearly funding gyrations, the low priority placed upon the project by some segments of the contractor team and NASA upper management, the many inexperienced people on the project and the pressures to contain costs), quality became a major concern for the project manager at NASA LeRC.

A deliberate philosophy was adopted to not decrease the test program and quality requirements. In fact, the net amount of testing was increased as the LeRC project team learned more about the potential weaknesses in the design. The quality assurance managers, Karl Reader at NASA and Al Little at Astro Space, deserve credit for ensuring quality in the electronic parts. They and their teams of parts engineers carefully reviewed, in a cooperative manner, the more than 400 requirements documents for non-heritage electronic parts. Schedule pressure to circumvent the process during the manufacturing phase was resisted by these two managers.

Test Program

The traditional spacecraft testing philosophy encompasses three levels of testing: the electronic box/mechanical component, the subsystem (e.g., multi-beam antenna and base band processor), and the spacecraft. The overall approach for ACTS was to protoflight test where the flight system was subjected to stress levels that were increased beyond mission requirements—to demonstrate design margins.

Because of the complexity of the ACTS payload, additional tests were also carried out to verify the design of this new switching and processing technology. Engineering models of critical boxes were built, tested, and integrated into a partial payload. This engineering model of the payload proved extremely valuable in verifying the complete payload design and its interfaces. It was also used with the master control station (MCS) as an early check on the MCS design, including the network control software.

The ACTS testing program, however, did not end there. At the completion of the spacecraft level testing, a fourth level of testing was implemented to test the ACTS spacecraft and MCS in an end-to-end series of communication tests with three of the user ground station types that would be used in in-orbit operations. This comprehensive system test lasted nearly three months and proved extremely valuable because it revealed a number of software problems that had not been uncovered during previous testing. Additional anomalies were uncovered in the high-speed command link to the base band processor. These software problems and anomalies would have been very difficult to diagnose in-orbit. After these communication tests, a special test was conducted to verify the RF continuity and switching for each of the 48 hopping spot beam locations. This spacecraft/ground system test was extremely valuable in further retiring technical risk and providing the confidence that ACTS was indeed ready to launch. Evidence of this fact is that on-demand digital services using the base band processor were successfully established the very first time they were attempted in orbit.

NASA LeRC ground segment team: (left to right) Ed Taylor, Tom Klucher, Rod Knight, Steve Mainger.

One of the more difficult items to test was the RF performance of the multibeam antenna. A completely new measurement capability, called a RF near-field facility, had to be built at Lockheed Martin to verify that the gain properties for 51 0.3° spot beams met their requirements. Approximately two weeks was allocated for these measurements. In the end, it took over three months to work out the test bugs in the new facility before NASA and Lockheed Martin engineers were satisfied with the results. There are many more stories about situations where quality was not sacrificed.

In any spacecraft program, there isn't enough time or money to verify every aspect of the design through testing. The review process, therefore, is a critical element in producing a quality product. During design reviews of all assemblies, subsystems, and the total system, it is necessary to make sure that there is adequate redundancy (in case of equipment failure) and that there are no single point failures of significance. After the design and manufacturing is complete, a complete as-built review of those aspects that are not verified by test should be made. One area that always falls into this category is verification of the correct polarity for electrical signals. This is extremely important for the attitude control system.

Thom Coney and Ron Bexten (not shown) of NASA led the ACTS system test in the high bay area at ASTRO SPACE.

On ACTS, the auto-track subsystem was a critical part of the attitude control for maintaining the pointing of spot beams within an accuracy of +/-0.025°. Using phasing information derived from RF signals transmitted to the spacecraft, the autotrack sends error signals to the spacecraft attitude control system that are used to maintain the correct pointing. There was no complete end-to-end test to determine whether or not the autotrack would provide the correct error signals. And, as it turned out, the normal review process had missed a design error in the autotrack that rendered its error signals useless. During one of these reviews, Charles Profera of Lockheed Martin questioned the correctness of the autotrack signal polarity. Since no end-to-end test verification was being made for this critical element, Mike Kavka, the Lockheed Martin program manager, instituted a late independent review of this subsystem. John Graebner, a senior and very experienced Lockheed

Martin engineer, performed the review and conducted a special test that confirmed the design error and necessitated a spacecraft change at the launch base. This illustrates the critical importance of reviews and the necessity that experienced engineers perform them. Inexperienced engineers who have not suffered through failures may not be motivated to perform a thorough check. Shame should be placed on the manager who presses his engineers to cut the review process short to save costs or time.

Setup for modifying the autotrack subsystem at the launch base prior to mating the spacecraft with the shuttle. The left portion of the photograph shows work platform extended out over the top of the spacecraft. Technicians used the platform to install shims behind the feed horn to correct signal phase. This delicate operation had to be performed extremely carefully in order not to cause spacecraft damage. The blue waveguide was for making autotrack RF phase measurements at the end of the Cleveland spot beam horn.

Checks and more checks are the recipe for producing a quality spacecraft. In-orbit, the only significant ACTS problem has been the wandering of the spot beams during a two-hour period each day. This deficiency, which is caused by a thermal distortion of the antenna subreflector surface (see Chapter 2, "Satellite Technology," for a description of the problem), was missed because of an inadequate thermal distortion analysis. Because of its complexity and cost, no direct thermal test had been performed to verify the thermal analysis. A better check on the thermal analysis should have been performed.

The ACTS System Has Proven Itself

ACTS has met all its objectives, has had no failures that reduced its capability to perform, and has exceeded its two-year operation requirement by over four years. All this has been accomplished without the use of redundant systems, except for a 20 GHz beacon transmitter that exhibited low power output and was turned off soon after launch. This beacon was activated again in December of 1999, and performed well. The low power was caused by losses in a mechanical waveguide switch that moved due to vibrations encountered during launch. Much of the credit for this success must be given to NASA, Lockheed Martin, COMSAT, and their subcontractors working as a committed team during the last few years before launch.

Keys to the spacecraft and master control station presented by Mike Kavka, Lockheed Martin, to NASA project manager Richard Gedney at buy-off.

During that time, in fact, the government/contractor team cooperated and focused on success. When a test revealed a problem (and there were many of them), not only was the problem identified and fixed, but the team strove to understand why the deficiency wasn't discovered sooner. Additional checks or reviews were then conducted to make sure the same type of deficiency did not occur elsewhere. A lot of credit for the success must go to Mike Kavka who, as the ACTS program manager at Lockheed Martin during this period, provided a lot of the necessary ingredients needed to form and maintain this committed government/contractor team.

CHAPTER 7

USING MARKET FORECASTING TO SHAPE ACTS TECHNOLOGY INVESTMENTS

Growth of Satellite Communications and Need for New Frequencies

INTELSAT ushered in the era of commercial communication satellite services with the launch of Early Bird (or Intelsat I) in April of 1965. Transoceanic cables were unable to keep up with the rapidly growing international demand for voice traffic, and incapable of transmitting the bandwidth of real-time television broadcasts. In 1965, the four transatlantic cables handled a total of only 328 circuits. With INTELSAT 1,300 additional circuits were added—revolutionizing transoceanic communication.

RCA inaugurated satellite communication services in the U.S. using transponders from Anik I, which was launched by Canada in November of 1972. Western Union launched Westar I in 1974 and used it for trunking, telex and TWX exchange services, point-to-point voice services, public messaging services, and data and facsimile services. It also provided video on a point-to-point and point-to-multipoint basis to a number of cities.

In 1977, voice and data were the major sources of communication satellite traffic, with video a distant third. In 1978, however, the growth of the cable TV systems using satellite distribution changed the market situation dramatically for satellite transponders. Almost overnight, a buyer's market changed to a seller's market. By 1979, twelve communication satellites were in orbit and providing commercial services. During this time period, communication satellites provided public switched telephony services, private-line voice and data services to large corporate users, and television-programming distribution to cable headends. TV programs are forwarded from the headend to the customer by cable. Personal computers were just being introduced. The Internet (or ARPANET, as it was called at the time), was the domain of scientists and researchers only. Fiber optics had not yet been installed to deliver long-haul telephone trunk services.

By the end of the 1970s, the growth of U.S. telecommunication (especially for video distribution to cable headends) was creating a greater demand for C-band (6/4 GHz) orbital positions than were available. Attention of the satellite industry shifted to Ku-band (14/12 GHz), and carriers began planning the launch of Ku-band communication satellites. In 1980, approximately 318 C- and Ku-band geostationary transponders (equivalent bandwidth of 36 MHz) were available for U.S. domestic, fixed satellite service (FSS) communication services.

Against this backdrop, the ACTS program was formulated in the late 1970s and the early 1980s. The primary objective for the ACTS program was to develop technology for commercial communication.

One of the main arguments supporting NASA's development of ACTS technology in the late 1970s was that the geostationary C- and Ku-band frequencies would become saturated, prompting the use of Ka-band, spot beam satellite systems. As part of the ACTS program formulation, market studies were performed in 1978 to better understand the direction of telecommunication growth and the role that communication satellites would play in handling some of this traffic. In addition, the studies were designed to determine the time at which the C- and Ku-band frequencies would become saturated. In this chapter, these market studies will be reviewed and analyzed to determine which market aspects were predicted correctly or incorrectly, and to illustrate the difficulty in making forecasts. These market studies were of fundamental importance because they convinced the ACTS sponsors that technology development needed to go forward.

Even as late as 1993, when ACTS was launched, there were experts (James E. Pike, head of space policy at the Federation of American Scientists [124]; Chris Dixon, senior VP at Paine Webber [124]) who still forecasted that the Ka-band would not be needed. With the explosive growth in the data communication market in the late 1990s, however, no one now disputes the fact that additional satellite capacity and the use of additional frequency bands is needed.

By 1997, there were approximately 980 C- and Ku-band transponders in-orbit. All the usable C- and Ku-band geostationary orbit slots for U.S. domestic communication have essentially been assigned (in-orbit satellites may not presently exist in all assigned slots).[1] The situation in 1999 is, therefore, that the use of the C- and Ku-band frequencies has reached a saturation level for satellite FSS services in the United States, as well as other locations around the world. It should also be noted that all the available Ku-band broadcast satellite service (BSS) slots in the U.S. and Europe have also been filled—although TV broadcasting service to homes was not considered as part of ACTS.

Initial Market Studies

In 1978, NASA awarded two contracts to common carriers—Western Union (WU) Telegraph Company and U.S. Telephone and Telegraph Company, a subsidiary of International Telephone and Telegraph (ITT)—for market studies [125, 126]. The emphasis of these studies was to provide estimates of future satellite-addressable traffic—that is, forecast telecommunications traffic that

---

1. There are some western U.S. Ku-band orbit slots that are not assigned because the significant Ku-band rain attenuation in the southeastern United States makes it difficult to provide high availability service from those slots.

209

could be carried competitively by satellite. The satellite-addressable traffic took economic competition with terrestrial delivery systems into consideration.

These initial market studies predicted that the demand for telecommunications in general, and for satellite communications in particular, would increase substantially over the next two decades. The rapid growth taking place in U.S. domestic voice, data, and video traffic was forecast to lead to a five-fold increase in traffic capable of being carried by satellites in the early 1990s. A combination of these market projections and communication satellite license filings with the FCC portended saturation in the capacity of the North American orbital arc using the C- and Ku-frequency bands by the early 1990s.

Two years later, in 1980, these market studies were updated. The point of saturation was found to be approaching even more rapidly than envisioned just two years earlier. In these two years, the demand estimate had virtually doubled, and the date of exhaustion on the conventional C- and Ku-band U.S. domestic communication satellites moved closer by five years [127].

1983 Market Assessment Update

After 1980, significant changes took place in the satellite communication industry. These changes included: technology advances, a wave of new satellite service providers, a rapid increase in the demand for cable TV channels, and the development of fiber optics. In response to these and other changes, NASA again updated the satellite traffic forecasts.

Three fixed service telecommunications demand assessment studies were completed in 1983 [128, 129, 130]. The studies provided forecasts of the total U.S. domestic demand from 1980 to 2000, for voice, data, and video services. The portion that is technically and economically suitable for transmission by satellite systems—both large trunking systems and customer premises services (CPS) systems—was also estimated. This portion was termed satellite-addressable traffic. It is important to emphasize that the satellite-addressable traffic is not an estimate of what will be captured by satellite systems, but is an estimate of what could be captured by satellite system operators. It is the amount of traffic that is potentially viable for transmission over satellite systems.

Each study contractor (WU & U.S. ITT) conducted its study entirely independent of each other. Independent approaches to the forecasts were taken and different ground rules and assumptions were employed. These studies focused on identifying the potential demand for customer-premises-type satellite communication systems providing services to terminals on the user's premises—possibly shared with other local users.

The third study (by WU) quantified the demand potential for large trunk-type satellite system applications suitable for carrying traffic among the 313 standard metropolitan statistical areas (SMSA). At the time of the studies, satellites were beginning to provide considerable trunking services.

In order to provide the ACTS program with a single set of forecasts, a NASA synthesis of the above studies was conducted and reported [131]. Thirty-five services were considered: nine voice, seventeen data, and nine video services. These services are described here and listed in the table entitled "Telecommunications Services Forecast in 1983." [131] All quotes are from the study report in reference 131.

Voice Category

Besides the normal long-distance terrestrial telephone calls, this category included the following:

- The long-distance component of mobile radio traffic carried by the switched network (As we now know this is important today because of the large increase in cellular phones.)

- The distribution of programming for National Public Radio, commercial radio networks, and special events, and

- The distribution of high-quality music for cable TV

Data Category

Data transmissions were organized into two types of computer services (low-speed terminal/CPU interactions and high-speed CPU/CPU interactions) and message-type services.

The low-speed terminal/CPU services were envisioned to occur at speeds of 56 Kbps or less. The inquiry/response service was for terminal operations of a more urgent nature such as airline reservation systems and stock quotations. The videotex/teletext subcategory was thought to be an umbrella service covering a variety of interactive and non-interactive consumer information services displayed on the home video screen. Today we would relate these services to those being performed by the Internet.

The CPU/CPU interactive services were primarily thought to be low-data-rate with some transmission speeds in the range of 56 Kbps to 1.544 Mbps (T1). Other data services included in the market forecast were:

- data transfer: The transfer of information from one storage bank to another

211

- U.S. Postal Service electronic mail switching system (USPS EMSS): The volume of mail transferred electronically over the USPS systems

- mailbox: Messages that are stored in a central computer which the recipient accesses at his convenience

- administrative message traffic: Generally short, person-to-person messages, usually of an intracompany nature

- mailgram: Similar to the USPS EMSS service, but performed by companies

Telecommunications Services Forecast in 1983 [131]

VOICE	DATA	VIDEO
VOICE SERVICES	COMPUTER	BROADCAST VIDEO
MTS, Residential	Terminal-to-CPU	Network, Commercial
MTS, Business and WATS	Data Entry	Network, Non-Commercial (PBS)
Private Line	Remote Job Entry	CATV
	Inquiry/Response	Occasional
Other Telephone & Radio	Timesharing	Educational
Mobile	Point-of-Sale	Public Service (Telemedicine)
Public Radio	Videotex	Recording Channel
Commercial and Religious	Telemonitoring	
Occasional		
CATV Music	CPU-to-CPU	
Recording	Data Transfer	
	Batch Processing	
	MESSAGE	VIDEOCONFERENCING
	USPS EMSS	One-Way
	Mailbox	Two-Way
	Administrative	Full Motion
	TWX/Telex	Limited Motion
	Facsimile	Fixed Frame
	Mailgram	
	Communicating Word Processing	
	Secure Voice	

- communicating word processors: Adds communication capability to a printer/keyboard or CRT-based word processing system. This allows the input to be prepared on one system and sent via communication links to another system for output, editing, or manipulation. The advantage to the user is the ability to transmit original-quality documents with format control similar to letter and memo correspondence.

We refer to these services today as either Email or file transfer protocol (FTP) performed using the Internet, intranets, or virtual private networks (VPN). Surprisingly, these studies were quite visionary considering that the PC and IP services via the Internet were new concepts at that time.

Video Category

The nine video services used in the study are quite familiar. It should be noted that the CATV subcategory included low-to-medium powered direct broadcast satellites (DBS) operating in the fixed services frequency bands. PrimeStar, of course, is doing this today. Most significantly, the forecasting did not envision the amount of image, video, and audio information that may soon be provided over the Internet.

The recording channel was a pay service providing video material for home recording, generally during off-peak hours. This service has been superseded by the local video rental store, but will be eventually introduced using hard disk drive video recorders. In fact, TiVo and others are working on developing a similar video-on-demand service. It is just another example of an idea that can't be implemented until there is sufficient technology advancement. In this case, the necessary technology advancements are the development of low-cost, hard disk drives and transmission capability.

Methodology

The prime objective of the WU and U.S. ITT studies was to quantify the amount of communication traffic that could potentially be transmitted over satellite systems. This is the portion of the total U.S.-generated communication traffic that, due to its system economics and technical and user characteristics, would cause satellite systems to be the preferred transmission means. The satellite-addressable traffic was derived by first estimating the total domestic U.S. traffic for 1980, excluding local non-toll traffic. This total traffic was then forecast for the years 1990 and 2000 using the contractor's proven methodologies, which included interviews with the IT managers of many corporations. The intra-SMSA toll traffic component was then removed (the assumption being that this was too short a distance for potential satellite traffic) to isolate the net long haul (NLH) traffic. The NLH traffic forecast then formed the basis for deriving the satellite-addressable market. A certain amount of traffic is not suitable for satellite transmission due to time delay,

213

somewhat lower availability levels, and other factors. These factors were taken into account.

In developing the forecasts, the voice, data, and video traffic estimates were first derived in their natural source units, such as number of messages for voice, bits/year for data, and channels for video. Service usage statistics such as message length, overhead, peak-hour factors, and transmission efficiencies were then considered to give an estimate of the equivalent transmission capacity required to carry the traffic—voice in peak-hour half circuits, data in peak-hour megabits/second, and video in numbers of peak-hour channels. To convert these capacity requirements to transponder requirements, estimates were first made of factors expected to influence transponder throughput capabilities. These included the likely coding advancements permitting the reduction of required bit rate/digital voice channel, multiplexing advancements permitting greater data rates/unit bandwidth, and video compression improvements. For the CPS service, the transmissions were considered to be all-digital.

Impact of Fiber Optics

Although fiber optic transmission technology was considered by both WU and U.S. ITT as a terrestrial competitor of satellite systems, rapid advances in this field after the studies were completed made fiber optic systems more competitive with satellite systems than originally accounted for in the studies. The first major long-haul fiber cable was installed between New York and Washington, D.C., in 1983. In 1986, the U.S. interexchange carriers installed 34,000 miles of fiber cable, followed by 64,000 miles in 1987.

Because of fiber optics, the satellite transmission companies saw a large reduction in demand. For instance, in the early 1980s, trunking of bundled analog voice circuits using companded single sideband techniques was being carried over satellites. By the latter part of the 1980s, no voice-trunking traffic was being carried by satellites. In fact, the large amount of trunking traffic that was forecasted to be provided by satellites by these 1983 studies has never materialized because of fiber optics. For this very reason, many thought that the ACTS technology would not be needed.

On-Demand, Integrated Services

One very important finding of the forecast studies was that business users desired on-demand, integrated services (as opposed to permanently leased lines), even though these type of services were not being offered at the time. As a result of this forecast, the ACTS system was configured to provide such services. When the satellite was launched in 1993, it provided the type of new services that users expected.

Very Small Aperture Terminals (VSAT)

In late 1980, Satellite Business Systems (SBS) launched the first commercial Ku-band satellite to provide a full spectrum of communication services aimed primarily at business users. The SBS system employed large 18- and 25-foot earth stations, located directly on the customer's premises, that provided all-digital, fully integrated voice, data, and image transmission capability. These earth stations proved to be too costly to manufacture and install because of their size. Generally, only the larger corporate users could afford their installation. In the mid-1980s, the emergence of a small class of earth stations (at or less than 8 feet in diameter), called very small aperture terminals (VSAT), took place. Such terminals were better suited for location at a customer's business site. It was then recognized that future development of earth stations for the ACTS BBP should be of the VSAT type, with antenna apertures of 4 feet. At this size, installation costs would be greatly reduced.

Forecasts

The table entitled "Total U.S. Domestic Telecommunications Demand" presents the forecast for the total U.S. domestic (excluding non-toll) peak-hour demand or capacity requirement for voice, video, and data. These results are discussed next.

Voice Category

The business-oriented voice services appear to exceed the residential requirement by an order of magnitude—but this merely reflects the fact that the peak hour for each service type occurs at a different time of day, with the business busy hour dominating. The total voice demand, as measured in half circuits, grows by a factor of 6.75 from 1980 to 2000 or an average annual growth rate of 10% over the 20-year period.

Data Category

The data forecasts are presented for each of the 17 data services, in terms of peak-hour megabits/second. Two factors influencing the forecasted data traffic growth are the underlying growth in demand and improvements in efficiency of transmission. Although the basic demand grew by a factor of 15 (about 14% per year) over the period from 1980 to 2000, the dominating

215

influence of transmission efficiency improvements caused the transmission capacity requirement (expressed in peak-hour megabits/second), to exhibit a peak and then decline somewhat beyond 1990. Remember that in 1980, most data traffic was transmitted over the circuit-switched terrestrial telephony system. The forecaster correctly assumed that by 2000, high-speed digital data links would exist where multiplexing advancements would permit greater data rates per unit bandwidth.

As these figures show, however, the data market was underestimated. In recent years—due to the Internet and business needs—data traffic has increased at a rate of 20 to 30% per year. Remarkably, the forecast included Internet type traffic but missed the magnitude of Web pages downloaded to Internet users.

Video Category

The demand for broadcast video channels was forecasted to grow at a rate of about 11% per year during the 1980s and then level off to about 4% per year during the 1990s. The forecast included broadcast of TV channels by satellites directly to home receivers. This was correct, since PrimeStar had a service in the 1990s that used the FSS Ku-band with a 30-inch terminal for this purpose. In 1999, Hughes purchased PrimeStar with the plan that all the PrimeStar subscribers (2,300,000 in February of 1999) would eventually be converted to the Ku-band, DBS frequency band, using an 18-inch terminal by the year 2000.

Total U.S. Domestic Telecommunications Demand

Traffic Type	1980	1990	2000
Voice (Peak Hour, 103 Half Circuits)			
MTS Residential	89	215	473
MTS Business and WATS	1616	4223	9842
Private Line	870	2765	7000
Other	4	43	129
Total Voice	2575	7246	17444
Data (Peak Hour Mbps)			
Data Entry	35446	37525	30918
Remote Job Entry	5668	5794	2863

Total U.S. Domestic Telecommunications Demand (continued)

Traffic Type	1980	1990	2000
Inquiry/Response	7526	7959	5083
Timesharing	4307	1507	736
Point-of-Sale	355	2975	2398
Videotex	3	969	931
Telemonitoring	1	4	16
Data Transfer	108	165	590
Batch Processing	1263	324	306
USPS EMSS	31	143	244
Mailbox	12	121	150
Administrative	4630	11575	17895
TWX/Telex	6	1	1
Facsimile	443	945	1120
Mailgram	1	1	1
Comm. Word Processor	35	115	227
Secure Voice	1	35	201
Total Data	59836	70158	63679

Video (Peak Hour, No. Of Channels)	1980	1990	2000
Broadcast	57	158	233
One Way Videoconferencing	1	112	225
Two Way Videoconferencing	2	1859	8000
Total Video	60	2129	8458

Note: Videoconferencing contains a wide mix of channel speeds.

217

Satellite-Addressable Traffic

Service pricing was developed for both satellite and terrestrial transmissions to arrive at the volume of traffic that could potentially be captured by satellite system operators. The traffic predictions are given in the following table.

The customer-premises service was assumed to be for business traffic only because of the ground terminal cost. Today, VSAT is the current nomenclature use for CPS terminals.

Summary of Voice, Data, and Video Satellite-Addressable Demand (Equivalent 36 MHz Transportation)

Segment	Traffic Type	1980	1990	2000
Overall				
	Voice	310	641	1594
	Data	33	254	518
	Video	60	250	312
	Total	403	1145	2424
Trunking Segment				
	Voice	310	638	1578
	Data	0	17	41
	Video	60	240	295
	Total	370	895	1914
CPS Segment				
	Voice	0	3	16
	Data	33	237	477
	Video	0	10	17
	Total	33	250	510

In hindsight, the Year 2000 forecast can be compared to the actual U.S. FSS domestic satellite traffic in the year 1998. This comparison shows that essentially no voice-trunking traffic is carried by satellite—significantly less data traffic is carried and significantly more video is broadcasted. In terms of equivalent 36 MHz transponders, broadcast video in the year 2000 was projected to be 92, while the videoconferencing was estimated at 220. This estimate for broadcast video was underestimated by more than a factor of four while the demand for videoconferencing was over-estimated.

Despite these large discrepancies, the forecast was accurate in predicting saturation of the C- and Ku-band frequencies in the early 1990s. In order for satellites to carry a much larger portion of the ever-expanding data and video market, it was certain that spot beam satellites would be needed.

Final Assessment of 1983 Market Projections

The 1983 market forecast, designed to predict the satellite-addressable communications market years later, was visionary in that it:

- identified many new services that are similar to those provided by the Internet

- included an impact of all digital transmissions with statistical multiplexing, low-rate voice and efficient video compression

- realized the importance of integrating voice, video, and data onto a common channel and providing that service on-demand, and

- understood the increasing importance of video and high-speed data links

These visions were incorporated into ACTS, which made it a modern communication test bed when it was launched in 1993. Any communication traffic market study has difficulty quantitatively estimating the impact of technology (fiber optics, high-capacity personal computers, Internet, intranets, and digital-signal processing), and this 1983 market study was no exception. The significant under and over predictions were identified in the previous "Forecast" section. The forecast, however, fairly accurately predicted the approximate time the C- and Ku-band frequencies would become saturated for U.S. domestic FSS communications.

The market study was *essential* in formulating the general requirements for ACTS and providing motivation for its development. It provided the guideposts for the program.

Follow-on Market Assessment

Following the 1983 assessment, market studies were continuously performed to update the potential of the ACTS technology for commercial services. Special Ka-band, spot beam system configurations were considered for mesh network VSAT; aeronautical mobile; high-data-rate (gigabit/sec.); thin-route personal communication, and supervisory control and data acquisition services [132, 133]. These studies demonstrated the advantages of high-gain, spot beam satellites over conventional systems, in terms of circuits per unit spacecraft weight in orbit.

In 1994, a new round of industry market and system studies completed by Space Systems Loral and Booz-Allen & Hamilton [134-136] assessed the capability of new technology satellites, like ACTS, to provide services. Using these results, Grady Stevens [137] performed additional analyses to determine those services that new technology satellites could provide at less cost than terrestrial communications. Stevens concluded the following:

> (Satellite) on-demand...service has potential for being competitive. By channel-sharing among a sufficient number of users, the pro-rated space segment charge is reduced to insignificance. The net user cost is essentially the amortization or lease charge for the earth terminals. Consequently, it is essential that earth stations be small and inexpensive for these applications. Optimization procedures are attempted to establish the minimum possible costs. Given the cost data and models in hand, this process suggests earth terminals on the order of 0.5 meter are needed. Correspondingly, large and complex satellite antennas, on the order of 7 meters at Ka-band, would be required.

> The prime application is viewed as multimedia communications within VSAT-like networks, which is expected to be made prevalent with the common use of Microsoft Windows and OS/2 operating systems. Conventional VSAT technology cannot efficiently process such communications and we suggest the New Technology Satellites as an alternative.

Further studies have proved this to be true. Reference 138 estimates the satellite transmission cost for two-way, Internet service for consumers as $12 per month. The satellite hub and ISP operation, service provider, and subsidy for the terminal make up the largest portion of the total service price which is estimated to be $40 per month. Large consumer electronic companies such as Motorola, Thompson, Hughes Networks, Gilat, and Nortel are now addressing the issue of producing a two-way satellite terminal for a suitable consumer price under $600.00. Assuming the necessary MMIC advancement for the Ka-band uplink transmitter, such a price is thought achievable in large enough production quantities. Amortizing this terminal price over a reasonable period should result in a competitive Internet service such as being planned by iSKY.

Internet and multimedia services over satellite show signs of explosive growth as satellite operators reported their statistics in 1998 [139]. During 1998, existing operators took in an estimated $104 million in U.S. revenue from broadband satellite services, which represents 111% growth according to Richard Greco, president of Loral Orion Europe. Greco quotes a Frost & Sullivan prediction of compound 97% annual growth for broadband satellites. This would take the market to $1.5 billion by 2002. Intelsat CEO and director general Conny Kullman says that Internet services generated 7% of Intelsat revenues in 1998, and will contribute 14-29% in 1999. This makes it the fastest developing service in Intelsat's history.

The action seems to be focused on existing operators/satellites rather than the Ka-band hopefuls. The recent introduction of hybrid Ka/Ku services by SES Astra, and the advent of construction of Hughes' SpaceWay, Lockheed Martin's Astrolink, and iSKY (formerly KaStar) indicate that Ka-band satellites are just around the corner.

A recent consulting study [139] also predicts large increases in demand for service tied to Internet growth. "Satellites pose a formidable threat to cable modems for delivery of global broadband access," concludes a new report from U.S.-based Pioneer Consulting. Pioneer says that satellites will be second to cable modems in terms of total subscribers, but that they will gain a larger revenue stream because they can address residential and business customers equally well. As predicted by the ACTS studies, the broadband market seems to be coming of age for satellites.

Global Broadband Access Subscriber Summary (in millions) U.S.-based Pioneer Consulting [139]

Access	1998	2001	2004	2007
LDMS	0.00	0.85	4.00	11.16
Satellite	0.03	0.65	10.49	39.62
Cable Modem	0.57	8.54	26.86	46.78
XDSL	0.20	5.32	21.45	38.49
Total	0.80	15.36	62.80	136.05

SUCCESSORS TO ACTS

Satellites are emerging as an ideal medium for a host of new global communication services. NASA's Advanced Communications Technology Satellite has provided an in-orbit test bed for pioneering, evaluating, and validating some of the cutting edge communication technology needed to implement these new services for the twenty-first century. The development and in-orbit testing of ACTS has helped to herald a new advance for the role of satellites in the information revolution. ACTS pioneered four key technology breakthroughs: on-board circuit switching and processing, dynamic hopping spot beams, very wideband transponders, and use of the Ka-band portion of the frequency spectrum. The advanced switching technologies provided ACTS users with bit rates on-demand. The multiple spot beam technology permitted a high degree of frequency re-use and small terminals for direct user-to-user (mesh) interconnectivity. With a high degree of frequency reuse, the capacity for a single satellite in orbit can be much greater, with a resultant reduction in service cost. Trials with the ACTS test network have validated on-demand, high-bandwidth, integrated voice, video, and data applications that can potentially be provided at a service price that is competitive with terrestrial communications. In today's deregulated environment, the reduction in service cost is crucial if satellites are to remain competitive.

Comments from Industry

Once launched and working in-orbit, ACTS began to gain recognition as a significant step forward in satellite communication technology and a valuable test bed for new communication applications.

Motorola executives have said ACTS was key to the development of onboard processing for Iridium satellites [140]. Motorola developed the base band processor (BBP) for ACTS—perhaps the most complex communication payload ever contemplated for a satellite in the 1980s. Motorola's intimate knowledge of the ACTS BBP technology, coupled with their position as a world leader in communication, led to their vision of Iridium. Motorola's participation in the ACTS program helped it to pioneer the commercial use of a switchboard in the sky. Iridium, relying on onboard switching to route calls between its 66 satellites and the Earth, uses payload architecture with a direct heritage to the ACTS BBP technology [141]. Keith Warble, Motorola vice president of the Strategic Electronics office, commented, "Even though ACTS concepts were conceived in the late 1970s, it is still the pathfinder mission for high-data-rate wireless communication systems [141]."

The Hughes Spaceway system (one of the new Ka-band systems filed with the FCC) is designed to provide switched bandwidth-on-demand to end users with ultra small, inexpensive terminals. The first Spaceway is scheduled to be

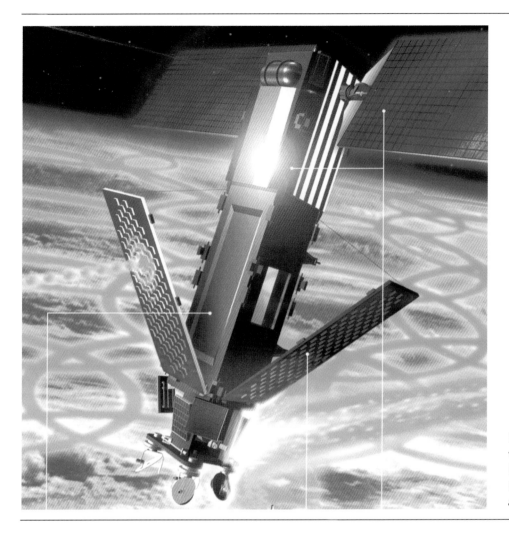

Iridium Satellite
with L-band Spot
Beams and Ka-
Band Intersatellite
and Gateway Links

launched in 2002 to provide multimedia services to North America. Steve Dorfman, president of Hughes Telecommunications, noted in a 1994 letter to
NASA administrator Dan Goldin [142]: "The Spaceway system builds on the
NASA ACTS technology concept, including onboard switching/processing
and an innovated 48-spot beam satellite architecture...". He further commented, "NASA has always played the role of champion for satellite communications." Ed Fitzpatrick, vice president of Spaceway, commented in another
letter to NASA [143]: "The NASA ACTS program has been valuable to Hughes
Spaceway as a pathfinder for implementation of Ka-band communication systems." At Satellite 97, Jerald Farrell, president of Hughes Communications,
commented that "Processor-based satellites, with their capability to transmit
wide bandwidth data on demand, is where the industry is going [144]."

225

Edward Tuck, founder of Calling Communications [later called Teledesic] noted, "It's been a huge advantage having ACTS up there. It brought a lot of credibility to what we're doing [145]." Teledesic vice president of Engineering, David Patterson, further commented that "We look at ACTS as a proof of concept... ACTS helped prove that Ka-band could be used for reliable, high-speed communications [145]." Teledesic is one of the few systems that are planning to use hopping spot beams as opposed to fixed beams. The hopping spot beams that were developed on ACTS allow Teledesic to adapt each satellite's communication capacity to efficiently meet geographically varying traffic.

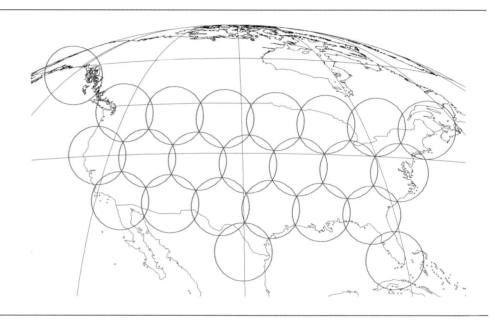

U.S. spot beam coverage provided by Hughes' Spaceway geostationary satellite. Each beam transmits and receives two separate polarized signals. For this reason the number of beams are double that shown.

Curtis Gray, a WorldCom vice president, observed, "The Advanced Communications Technology Satellite (ACTS) designed by NASA has revolutionized high-speed satellite communications. Due to this innovative design, satellites will continue to play a major role in building the Global Information Infrastructure (GII) [146]."

AT&T manager Paul Moravek explained that "...the Advanced Communications Technology Satellite (ACTS) has moved satellite communication to a new plateau, promising communication satellite data rates previously only possible through terrestrial communication. This is significant, as it provides the potential to shrink the world even further, and enables new applications such as multimedia broadcasting to remote locations not economically served through existing technologies [147]."

Lockheed Martin's Russell McFall, president of Astro Space Commercial, observed, "By pioneering the use of Ka-band, ACTS has achieved its goal of

226

maintaining United States preeminence in satellite communication, and has spurred a global industry that will allow universal and economic access to the NII/GII for people all over the world [148]." Lockheed Martin has filed to provide a global network of Ka-band, spot beam satellites called AstroLink. AstroLink incorporates much of the ACTS technology, including the using of adaptive rain fade compensation, to reduce onboard processing power. Adaptive rain fade compensation is a unique technology developed by ACTS. The first AstroLink is scheduled to start delivering service in 2003.

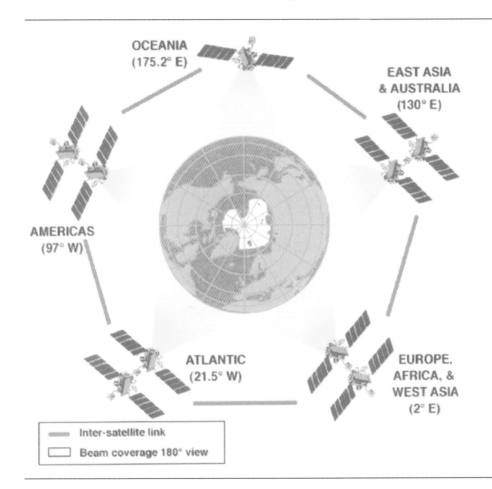

OCEANIA (175.2° E)

EAST ASIA & AUSTRALIA (130° E)

AMERICAS (97° W)

ATLANTIC (21.5° W)

EUROPE, AFRICA, & WEST ASIA (2° E)

Inter-satellite link

Beam coverage 180° view

Astrolink Satellite Constellation Providing Global Connectivity at Geostationary Orbit

John Evans, President of COMSAT Labs, commented that "The flexibility provided by ACTS technology permits efficient utilization of space segment resources in meeting a wide diversity of telecommunications needs. ACTS provides the cleanest and most compelling demonstration of the key role future satellite systems will play in the planned NII/GII [149]."

Jeremy Rose, a noted British communications consultant, wrote in a recent paper: "In the U.S.A., the Advanced Communications Technology Satellite (ACTS) is being used to prove a number of technology concepts, including scanning beams and onboard processing, and has been one of the most important drivers for the commercial Ka-band satellite proposals currently under development [150]."

Walt Morgan, a respected communication consultant for over 30 years, summed it up when he noted: "Those 14 applications [FCC applications for Ka-band satellites]—it's basically a vindication of NASA's belief in the technology [151]."

John Flannery, executive director of the United States Space Foundation, said, "Your organizations' role in the successful development and commercialization of the Advanced Communications Technology Satellite is being recognized as an excellent example of taking space technology and applying it to the betterment of life here on our home planet."

The United States Space Foundation, in its yearly symposium in Colorado Springs, recognizes space technologies that have had a significant impact on the commercial world and contributed to improving human life by inducting them into its Space Technology Hall of Fame. ACTS, NASA, and the contractor team were inducted into the Hall of Fame on April 3, 1997. See Appendix A for a list of the inductees.

Communication Revolution

In the last few years, the world has been experiencing an unprecedented demand for communication services. The rapid growth of the Internet is just one of the most visible manifestations of a whole host of new and innovative user services. It seems that there is an unquenchable public thirst for communication. On the corporate level, desktop computers are ubiquitous and often are being replaced by high-end workstations. More frequent video conferencing sessions punctuate the schedules of worker and managers alike. A key element for corporate communication is networking, both within and between organizations. Since business is conducted on a global scale, networks must provide global interconnectivity. On the personal level, we are surrounded by a plethora of affordable gadgets that allow us to communicate with others: Email, voice mail, facsimile, computers, cellular telephones, pagers, direct broadcast TV, Internet browsers, and desktop video conferencing.

The aggregated demand (see Chapter 7, "Using Market Forecasting to Shape ACTS Technology Investments," for the latest forecasts) for both corporate and individual communication services is pushing networks to run at higher and higher data rates. In many cases, broadband satellites will be able to best serve this demand.

Technology Hall of Fame Inductees: (left to right) Robert Lawton—1st Astro Space Program Manager, Richard Gedney, Charles Schmidt, Mike Kavka—3rd Astro Space Program Manager, Frank Gargione and George Beck, who except for Gedney were all part of the Astro Space satellite team.

Technology Hall of Fame Inductees: (left to right) Thom Coney, Mike Zernic, Robert Bauer, Gerald Barna, Richard Reinhart, Richard Gedney, and Rod Knight. All were part of the NASA team.

Explosion in Ka-band Filings

In the mid-1990s, it was clear that the existing C- and Ku-band frequency being used by satellite operators had insufficient spectrum to provide for the projected new interactive, high-data-rate services. To satisfy this demand, satellite manufactures and service providers began to turn to Ka-band, multiple spot beam, switching and processing satellites using some or all the elements of the technology pioneered by ACTS. As a result, there was a land rush for Ka-band spectrum. In 1995, U.S. satellite developers laid claim to Ka-band spectrum in response to the FCC's deadline for accepting the first round of satellite filings. Filings to the ITU from other countries around the world soon followed the U.S. initiative. As a result, a whole new generation of sophisticated global communication satellite systems has been proposed. Developers plan

229

to invest well over $30 B to make Ka-band the home of a wide range of high-speed, two-way, digital communication services within the next ten years [152]. Most of these systems will be global in scope and will provide services directly to businesses and consumers on their premises.

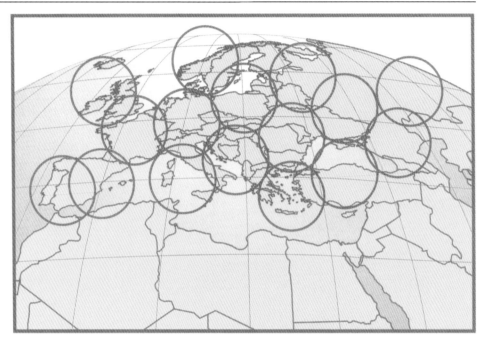

SES ASTRA 1K Coverage. The 16 Ka-band spot beams are 30 GHz uplink transmissions from the user terminals. The 16 spot beams are all fed into a single downlink beam at 20 GHz. The forward link back to the users is provided by a single Ku-band beam. The user terminal provides interactive services using a 60 cm antenna and 0.5 watt HPA.

Appendix B details the various geostationary Ka-band satellite systems that have been proposed and filed with the ITU and/or the FCC [151] through mid-1997. The ACTS technologies of onboard switching and processing, hopping spot beams, and wide bandwidth transponders are being used by a large number of these proposed systems. This list, which was prepared by Walt Morgan in 1997, does not include the second-round filers to the FCC. It should be noted that since the original filings, some systems have been withdrawn (most notably AT&T's VoiceSpan system—not included in the list), while others have been modified.

In the United States alone, 17 different communication satellite systems have filed with the Federal Communications Commission (FCC) to provide both domestic and international services from GEO, for a total of 92 Ka-band communication satellites. As of late 1999, the FCC had granted licenses for 73 of these satellites.

The use of multiple spot beams and onboard switching and processing is not confined to Ka-band geosynchronous satellites. Additional filings have been made to the FCC for use of Ka-band for non-geostationary fixed services and feeder links for big LEO systems. The most notable of these non-geostationary systems include Motorola's now operational, 66-satellite Iridium system (featuring Ka-band cross and feeder links) and the proposed Teledesic's 288-satellite system.

Of course, not every proposed system will be funded, built, and supported by the market. Historically, not all filings have turned into reality. For example, of the 1984 filings for C- and Ku-band satellites, more than half were abandoned. In 1984, the market for C- and Ku-band satellites was better understood than that for Ka-band today. However, financing is probably more readily obtainable today. Even if only a small fraction of these proposed systems obtains orbit, it is clear that there is an enormous growth ahead in satellite communication.

As of early 2000, seven commercial satellite systems with Ka-band transponders have been, or are being, constructed: these are KoreaSat-3, Societe Europeenne des Satellites (SES)–Astra 1H and 1K, Lockheed Martin's AstroLink, Hughes' Spaceway, iSKY (formerly KaStar) and Loral CyberStar. Astra 1H and KoreaSat-3 were launched in 1999. Astra 1K, along with the first of several iSKY spacecraft, will be launched in 2001. The first of several Astrolink, CyberStar, and Spaceway spacecrafts are to be launched in 2002. SES's 1H and 1K satellites provide Ka-band, bent-pipe, spot beam uplinks from the user and a forward link back using Ku-band DTH transmissions [153]. They recently received the first Ka-band, two-way user terminals (actually a hybrid Ka/Ku terminal) for commercial services from Nortel—a major milestone. KoreaSat will initially concentrate on tele-educational services [154], while both Astrolink and Spaceway are designed to initially serve the business market, with Spaceway potentially providing two-way consumer Internet services. iSKY will concentrate on Internet services.

As discussed in the previous chapter, the Internet/multimedia service revenues for 1998 and 1999 grew at a rate of approximately 100%. Intelsat is currently providing Internet services in 80 countries and expects the revenues from these services to be 20% of its business in 1999. So far, the initial satellite broadband business has been satisfied by current assets of the service providers. As the satellite market continues to grow, there will be a natural evolution into the Ka-band arena. The initial approach was to take the multiple spot beam and onboard processing technology and try to apply it to *all* broadband markets. This has turned out to be a mistake. The natural evolution will be to analyze the market segments and determine the best ways that satellites can respond. With the large broadband market growth, the new ACTS-type technologies will certainly have a role to play.

Array antennas for ICO global communications satellites (in the foreground, one receive antenna is complete, with 127 radiating elements for forming multiple spot beams at L-band).

ACTS Technology Use at Other Frequencies

The concepts of multiple spot beams are equally applicable at frequencies other than the Ka-band. Mobile communication is a good example. A relatively small amount of frequency spectrum has been allocated in the L-band for satellite mobile communication. Such satellite systems as Iridium by Motorola; GlobalStar by Loral; ICO Global Communications by Hughes; Asia Cellular Satellite (ACeS) by Lockheed Martin; Asia Pacific Mobile Telecommunications Satellite (APMT) and Thuraya by Hughes; and Ellipso Mobile Communications Holdings by Boeing all use multiple spot beams with a high degree of L-band frequency reuse to achieve a large enough satellite capacity to make their systems feasible. Iridium and GlobalStar began operations in December of 1998[1] and the first quarter of 2000, respectively. Many of these multi-beam systems (Iridium, AceS, APMT, and Thuraya) have onboard processing. The Iridium's onboard, base band switching has already been discussed. ACeS, which was launched in February of 2000, has an onboard, transparent, digital signal processor that performs demultiplexing, switching and

---

1. As of March 24, 2000, Iridium was bankrupt and had ceased operating.

multiplexing functions that enable feeder link to spot beam routing on a per channel basis and a spot-to-spot routing on a TDMA slot basis.

SkyBridge, an 80-satellite, low-earth-orbit constellation by Alcatel, will provide fixed broadband services using the Ku-band. Each SkyBridge satellite provides up to 45 simultaneous spot beams. The use of Ku-band for this system is advantageous because rain fade at Ku-band is much less than at Ka-band.

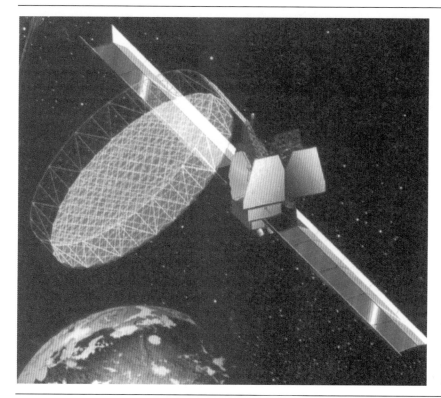

Thuraya (left) and Asia cellular satellite (next page) are two geostationary satellites which will provide handheld mobile services in the Middle East and Asia-Pacific, respectively. Each satellite forms hundreds of L-band spot beams, using 12 meter reflectors to obtain a high degree of frequency reuse.

Thuraya and ACeS are produced by Hughes and Lockheed Martin, respectively.

Concluding Remarks

The preceding statements by leaders of the satellite communication industry, the many Ka-band systems being proposed/constructed (see Appendix B), and the start of global operations by Iridium are all proof that ACTS achieved its mission of assisting with technologies needed by satellites in the new era of communication. We can now see that communication satellites are poised to take on new and expanding roles in the national and global information infrastructures.

The advanced technologies that have been flown and flight-tested with ACTS have greatly aided in providing the technology bridge to a whole new generation of twenty-first century communication satellites.

233

Asia Cellular Satellite

CHAPTER 9

THE ROLE OF GOVERNMENT IN TECHNOLOGY DEVELOPMENT

1987
SPACE TECHNOLOGY HALL OF FAME

Presented to

NASA Lewis Research Center

INNOVATORS

Advanced Communications Technology

Au F m Space Benefiting Man E rth

Many government-sponsored communication programs are undertaken to develop technology and/or create an environment which leads to increased private sector investment, and the formation of ventures which are in the public interest. In the U.S., government promotion of communication technology has a long heritage. For instance, in 1843, Congress appropriated funds for a demonstration of Samuel Morse's telegraph system between Baltimore, Maryland and Washington, D.C. Its successful completion and operation led to the commercial expansion of telegraph service throughout the United States [155]. Since World War II, the federal government has assumed a strong role in building the nation's science and technology capabilities. The legacy created by Dr. Vannevar Bush's Office of Scientific Research and Development changed the course of scientific and technological R&D in the U.S. by creating a large and increasingly active role for the federal government.

At the start of the space era, relatively little was known about the extreme environment of space. As described in Chapter 1, "Program Formulation," research and development carried out by NASA and the DOD in the late 1950s and the early 1960s served to reduce the technical risks associated with both the launch and operation of communication satellites. Following the 1960s, NASA's ATS program developed technology for commercial use and was highly beneficial to the industry. ATS developed political ambivalence during its later stages, however, and was phased out in 1973 because of budgetary pressures [1]. The ACTS program was initiated in the 1980s because of private industry's lack of resources for long term R&D. The ACTS program, although generally considered a success by most people, was attacked during its lifetime for being unnecessary. This chapter reviews the impact of recent government-sponsored technology for satellite communication, the criticism levied against federally sponsored developments, the current state of the satellite communication marketplace, and considerations for the role of government in the new century.

The Need for Proven Technology

In an April, 1998 article [156], Robert E. Berry, president of Space Systems/ Loral was asked: "In the early years, so much of any satellite was brand-new technology. You may never have a satellite where 100 percent of the components are proven, but how close are we?"

He replied:

> We're very close; in fact we are there. What we deliver... I can tell you that 100 percent of the components have been applied and flown. That is everywhere; all the commercial suppliers have that. Electric

propulsion is being applied for the first time in a commercial communications satellite. But in fact, it has established a history of space flight, so you can easily say it has been proven.

The commercial market does not embark on things where there is an unproven component. It's sort of like a bank making a loan. They never make a loan that is not sure to be paid back. Some of them don't get paid back, but when they make the loan, they were assured six ways to Sunday it was going to be paid back.

That's the way we are delivering the commercial satellites. Every component has a history; it has been proven and established.... In combination with the rest of the system it is a minimal risk approach.

That is why the commercial industry has grown so fast, and has a history of success and reliability that is so high.

In the world of fixed-price contracts with in-orbit incentives, there is little room for technical risk. The industry has continually made significant improvements. How, then, does technical innovation get implemented in commercial satellites?

The industry itself invests R&D funds that result in incremental steps—usually small or modest. These incremental evolutionary improvements added up over a long period of time have resulted in substantial gains in efficiency and the introduction of new services. During the late 1970s, INTELSAT (in particular) often forced technology improvements by specifying system performance requirements that could be met only by implementing new technology. Without a doubt, however, the U.S. government's (both NASA and DOD) sponsoring of satellite communication research resulted in many improvements. Government programs frequently tackled long-range, high-risk technology developments that led to technology innovations like the ACTS switching and processing technology.

NASA's and DOD's sponsorship of satellite communication technology helped to foster commercial growth and served as a catalyst to drive the initial development of the now burgeoning commercial satellite communications industry. This government investment has provided a rich technical heritage for many of the features of the commercial communication satellites built by U.S. companies to date. The accompanying table illustrates the government technology spin-off for recent commercial programs. Most of this government research is for its own mission-specific programs, but proves to have application in the commercial world.

Satellite Communications asked the following of Robert E. Berry [156]:

"Under Loral's ownership, the company has built more momentum in the commercial market, and less reliance on government business. Where is that trend headed?"

Government Technology Spin-Off

Commercial Satellite	Application	Technology	Prime Technology Sponsor			
			DOD Mission	NASA Mission	NASA Commercial	Commercial Industry
Iridium & GlobalStar	NGSO Mobile	L-Band Phased Arrays	Raytheon Radar Systems			
Iridium, Astrolink, SpaceWay, etc.	NGSO Mobile and GSO Broadband	Regenerative Payload with Baseband Switching	Milstar FLTSAT-7,8 UHF Follow-On		ACTS	
PanAmSat PAS-5 Galaxy VIIIi	GSO	Ion Propulsion for Station Keeping		SERT IAPS		
AstroLink, Spaceway, Teledesic, etc.	NGSO and GSO Broadband	EHF Spot Beam Antennas & RF Components	Milstar FLTSAT-7,8 UHF Follow-On		ACTS	
ACeS	GSO Mobile	12 M L-Band Reflector		TDRS		
DirecTV, EchoStar, etc.	BSS	Video Compression MPEG 1&2	√	√	√	√
Spaceway, Astrolink, ACeS, etc.	All Services	ASIC	√	√	√	√
	NGSO GSO	Radiation Hardening	√	√	√	√

His response indicates the value of government R&D:

> Well, we are happy to have government business and we will work hard to succeed in that. We have a relatively higher percentage of commercial activity than any other space manufacturer. They all have a significantly greater percentage of government activities than we do.
>
> For our government business, we focus on activities closely related to our skills in commercial business. The kinds of things that are called into play in commercial business are the ability to work with technology that is more certain than uncertain, with costs and deliveries that are more certain than uncertain.
>
> There is still a component in the government business—and it should be there—where space is put to work to try things that had never been accomplished before. There was a drive to advance national technology by undertaking things that, maybe, couldn't be done. And of course, it would always take longer, and cost more money, and sometimes it wouldn't work. That is still present in some of the national space endeavors. Less so than previously. But there is a need to stretch technology in the field of science and in the case of national security sometimes, to go for performance that may not have been achievable previously.

Not only is the government spin-off of technology very valuable, as Berry implied, but every NASA and DOD contract authorizes a percentage of the total funds for internal research and development (IR&D). Although government contracts usually yield low profits, the value of these guaranteed IR&D funds are immeasurable.

With the vast amounts of NASA and DOD space program R&D expenditures, there is no doubt that the government has played—and continues to play—a key role in keeping the U.S. commercial space industry vital and robust. The government's role has been—and continues to be—to sponsor long-term, high-risk technology.

The ACTS Approach

The ACTS program had a different approach than that of the U.S. government's normal R&D for communication satellites. Its prime purpose was to develop technology, not for a specific NASA mission but for use in providing commercial communication. This created two adverse reactions. Within NASA and the administration, there were detractors who thought that NASA should not be aiding the commercial industry and that it should only expend funds on government missions. Outside of NASA, there were some who thought that NASA's involvement would be detrimental rather than helpful,

239

and questioned whether the ACTS technology was even needed. In the late 1970s and the early 1980s, the satellite communication industry generally thought that ACTS technology should be developed and that the government should sponsor that development because it was beyond the capability of industry to fund during this time period. NASA's $50 M proof-of-concept laboratory development of ACTS technology from 1980-1983 was broadly supported and not challenged. The main issue revolved around whether or not it was necessary for the government to sponsor a flight program. The arguments for a flight program were:

- the need to counterbalance foreign government (Europe and Japan) flight program subsidies, and maintain U.S. leadership

- the service industry's perception that the technical risk was too great for incorporating the ACTS technology in a commercial satellite without a flight demonstration

The arguments against a flight program were:

- the winning contractor for ACTS would have an unfair competitive advantage

- the belief that the government is not capable of successfully guiding flight system development for commercial purposes

Each of these arguments will now be discussed in detail, to determine if they apply today. The findings from these discussions will be used to talk about the future role of government in technology development for satellite communication.

Foreign Competition

International competitiveness was a key issue in the early 1980s. Of course, this was prior to the fall of the Berlin Wall and the demise of USSR. Any issue with the competitiveness label could rise to the top of the national agenda. Commentators, as well as business and political leaders, warned that the U.S. was falling behind the Europeans and (especially) the Japanese in a broad range of industries [13]. In fact, it was possible for the ACTS advocates to show that the annual government funding for satellite communication by both the European Space Agency (ESA) and the Japanese Space Agency (NASDA) was much greater than by NASA (see figure in Chapter 1 on page page 8).

In the late 1990s, the argument that foreign suppliers would surpass the United States in the marketplace does not have nearly the weight it had in 1984—the first year that Congress authorized funds for the ACTS development. What has changed? The convergence of communications and computer technology has led to an exponential growth in communication services of all kinds. Communication has become an important aspect of business life, which is now being conducted on a global scale. Like the semiconductor industry, satellite communication companies serve global markets and have now formed global alliances that foster cooperation between companies from different countries.

A good example of this is Sematech, which was formed as a joint government/industry venture in 1988 to bolster the U.S. semiconductor industry's global competitiveness. At the time of its formation, Sematech had an annual budget of about $200 M, half provided by the U.S. government and half by U.S. chipmakers. Federal funding ceased in 1996. Sematech first accepted international participation in 1996, when five U.S. companies (along with Samsung Electronics and LG Semiconductor—both of South Korea), joined its research initiative on production of 300-millimeter silicon wafers—the base material for chips.

As recently reported in the Wall Street Journal [157]:

> Hyundai Electronics Industries of South Korea, Philips Electronics of the Netherlands, SGS-Thomson, a Franco-Italian concern; Siemens of Germany, and Taiwan Semiconductor Manufacturing—created in 1998 an international expansion of the Sematech partnership. This action is driven in part by the increasing expense of research and capital investment as chips become more complex.

The Wall Street Journal article went on to say:

> It shows perceptions have changed in the U.S. chip industry about foreign competitors. 'There has been a softening of the U.S. attitude,' said Howard Dicken, former executive at Motorola and now a chip-market analyst at ICE Corporation in Phoenix. Geographic boundaries are pretty well disappearing.

Similar alliances with foreign suppliers are occurring in the communication satellite industry. For example, Space Systems Loral has hired Alenia from Italy to assemble the GlobalStar satellites that will be used to offer worldwide mobile communications. Teledesic plans to have Matra Marconi build the spacecraft bus systems for its 288 satellites that will deliver broadband services around the world.

The fact is, the global marketplace is forcing companies from around the world to cooperate. This doesn't remove the need for government-sponsored R&D, since the economic health of any country will greatly depend on its technological capability. It means the U.S. government must pay more atten-

241

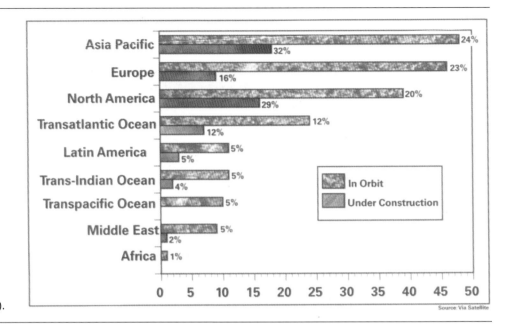

Region	In Orbit	Under Construction
Asia Pacific	24%	32%
Europe	23%	16%
North America	20%	29%
Transatlantic Ocean	12%	12%
Latin America	5%	5%
Trans-Indian Ocean	5%	4%
Transpacific Ocean	5%	
Middle East	5%	2%
Africa	1%	

Source: Via Satellite

Global market for geostationary communication satellites in 1999 (source: *Via Satellite*, July 1999).

tion to those areas where there are significant technology barriers. Such an area is semiconductor chips, where industry suppliers face fundamental technological barriers that could test their ability to innovate and grow at historic rates. According to a 1997 *Wall Street Journal* article [158], the chip industry is "lobbying to get a greater share of federal research spending" in addition to "forming consortia to work on specific problems as well as contribute tens of millions of dollars for university research."

Commercial Systems: Low Technical Risk and Flight-Proven Hardware

As stated by Robert Berry of Space Systems Loral, satellite service providers generally require that flight-proven components be used in their systems. Many of the ACTS technologies were such a departure from what had been previously flown that the service providers definitely felt a flight demonstration was necessary. This was especially true for the onboard base band switch. The switch had a large electronic component count, which caused the service providers to have serious doubts about its reliability. In addition, it used specially designed integrated circuit chips never before attempted for space flight. Such chips, now commonly called application-specific integrated circuits (ASIC)—the name didn't exist at the time of ACTS—are widely used in both terrestrial and satellite communication.

The developer of the base band switch for ACTS was Motorola, and that experience proved important for them. The October 1997 issue of *Iridium Today* [140] expressed the value of the ACTS development. It states:

> Motorola a participant in the ACTS program, developed the on-board switching system that led to its current use in Iridium satellites. Motorola's participation...helped it to pioneer the commercial use of a 'switchboard in the sky,' and to lead the revolution that will make tele-communications satellites cost-competitive with terrestrial applications. The Iridium satellite constellation is bringing to subscribers worldwide the prime technology concepts proven by the ACTS program: onboard switching for subscriber communications.

High technical risk should not be confused with financial and market risk. Private industry has demonstrated that once a new technology has been proven suitable for space application, it is then capable of incorporating it successfully. Although this isn't always true, it is true a high percentage of the time.

What has really changed is the perception of financial risk. Satellite projects, viewed as exotic and risky in the early 1980s when ACTS was being formulated, are now, in the late 1990s, experiencing a surge of interest from investors lured by their high earnings potential and capability to tap global markets.

Aviation Week & Space Technology recently cited an example [159]:

> When PanAmSat Corp. executives first came calling on Wall Street in 1993 looking for several hundred million dollars to expand their privately owned satellite network, the company's prospects were considered so risky that its bonds received a junk rating.
>
> Today, PanAmSat has expanded to a global network, signed customers to decade-long transponder leases, and achieved earnings margins of almost 80% before taxes and amortization. The company has worked its way up through bond ratings to investment grade—an amazing leap in five years by Wall Street standards.

The perceived market potential for satellite communication, as well as for other new technology, has changed and is much greater today than in the 1980s. The satellite industry and its investors recognize the important role satellites can play in effecting a truly global information superhighway. In under-developed countries, satellites can put a communication infrastructure in place much faster than fiber optics or other terrestrial systems—and at a fraction of the cost-per-customer. The market for satellite communications has moved from serving a few of the top 25 corporations and governments in the

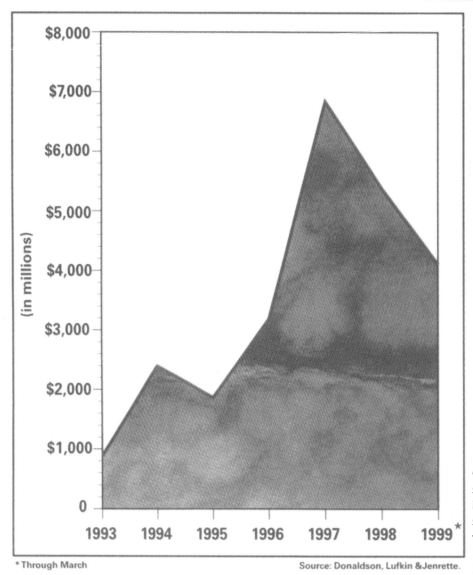

Capital raised in satellite services industry for new satellites and ventures (source: *Via Satellite*, July 1999).

* Through March   Source: Donaldson, Lufkin &Jenrette.

1980s to encompass all of us—not only in our business dealings, but also in our personal lives. Walter Morgan points out:

> The forthcoming Ka-band satellites will go further down the user pyramid to provide basic Internet, videophone, and other services to homes and small businesses. Some of the eventual applications for high-speed access are as invisible today as the Internet was in 1977. [160]

It is projected that the satellite industry will need at least $30 B in capital over the next 5 to 7 years to fund the various new space communication projects [159]. Investment bankers caution that only a fraction of the space proposals on the table will ever materialize. But financially viable ideas do have a better chance at winning funding, thanks to a booming stock market and a maturation of the space communication industry that has made investors more confident. This maturation has taken place by the introduction of a steady stream of successful technology improvements and innovations by the space industry. Government projects like ACTS, ATS, and MilStar can take a lot of credit for many of these technology improvements and innovations, especially the ones that were initially high-risk programs. However, it still remains generally true that a new innovation must be flight-proven before it will be incorporated by a commercial company.

Competition Between Satellite Manufacturers

Satellite manufacturers that opposed ACTS, did so on purely competitive grounds. In the commercial world, satellites are procured on a fixed-price basis with financial incentives for successful delivery and in-orbit operations. The top three major suppliers of commercial geostationary satellites in the world at this time (and at the time of ACTS) are Hughes Space and Communications, Lockheed Martin (the builder of ACTS), and Space Systems Loral. All three have always competed strongly against each other, with the result that their profit margins are not large. Any technical superiority that one company can gain over the other two can be a significant advantage to securing additional contracts.

Satellite Industry Subsegment	1996*	1997*	1998*	Change '96-'97	Change '97-'98
Satellite Manufacturing	$12.4	$15.9	$17.6	28%	10%
Prime Contractors	$8.3	$10.6	$11.7	28%	10%
Subcontractors	$4.1	$5.3	$5.9	29%	11%
Launch Industry	$6.9	$7.9	$7.0	14%	-12%
Launch Services	$4.2	$4.8	$4.3	14%	-11%
Manufacturing Subcontractors	$2.7	$3.1	$2.7	14%	-14%
Satellite Services	$15.8	$21.2	$26.2	34%	23%
Transponder Leasing	$5.2	$5.8	$6.1	11%	5%
Subscription/Retail Services	$10.6	$15.5	$20.1	46%	30%
Ground Equip. Manufacturing	$9.7	$12.5	$15.2	29%	22%
TOTAL	$44.8	$57.5	$65.9	28%	15%

*in billions $

Source: Satellite Industry Association

Global Satellite Revenues

245

An argument could be made that the winner of an experimental advanced technology flight system procurement (such as ACTS) would gain a significant competitive advantage for the development of future commercial systems. Since the ACTS procurement was a cost-plus-fee type of contract, the winner also had little financial risk. It is felt that this might have motivated Hughes to openly oppose the award of the ACTS contract. Very recently, Steven D. Dorfman, chairman of Hughes Space and Communications [161], said he favors NASA's long-term research and development that focuses on devices rather than satellite systems. Dorfman most likely feels that NASA's future development of entire satellite systems like ACTS will still create an undesirable distortion in the marketplace. The U.S. Congress also supports this view at the present time.

As it has turned out, no one company has gained a significant competitive edge because of ACTS. Many companies (Lockheed Martin, TRW, Motorola, EMS Technologies, Composite Optics, and COMSAT) performed the critical technology developments so that there was no technology monopoly created in a single company. The greatest achievement that ACTS has made is to convince people that ACTS-type technology can be successfully flown. In the year 2000, the actual ACTS technologies have been superseded by superior developments.

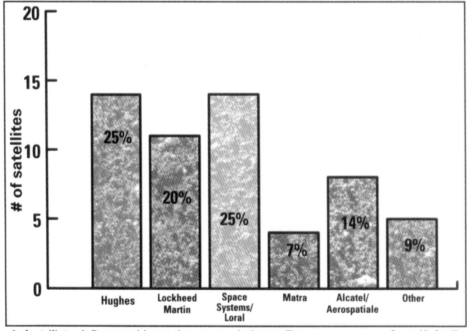

1998 satellite manufacturer market share – geostationary communication satellites under construction.

The Government's Capability to Guide Programs for Commercial Markets

The ACTS program had been formulated in concert with industry, using market and technology studies performed by the satellite communication industry. Based on the results of these studies, a substantial (approximately $50 M) proof-of-concept laboratory development of that technology was conducted from 1980–1983, with the full support of the satellite industry. There were other people, however, who maintained that the development being accomplished was not necessary and that the technology was not needed. They also believed that the Ka frequency band would not be needed in the next 20 to 30 years. These individuals believed that government was not capable of conducting programs for the benefit of commercial markets. It is a philosophical view that was also taken by many in the Republican administration during the 1980s and early 1990s.

William J. Broad perhaps best expressed this view in an article in the *New York Times* [162] on Tuesday, July 20, 1993. The following are excerpts from that article, which appeared just before the launch of ACTS in September of 1993:

> Private analysts say ACTS is a case study in federal myopia. They note that the craft was planned just as the rising attractiveness of fiber-optic cables on the ground began to bring much of the telecommunications industry crashing back to Earth. [This statement is true about fiber optics, but ACTS wasn't designed to provide services that competed with fiber-type trunking services.]
>
> More generally, many economists say federal officials lack the knowledge to predict what technologies will succeed in the marketplace and are never canny with taxpayer money, unlike entrepreneurs who risk their own. Such defects, they say, make federal industrial policies all too prone to producing white elephants.
>
> 'It's lemon socialism,' said John E. Pike, head of space policy at the Federation of American Scientists, a private group in Washington. With ACTS, the government is funding stuff the market has rejected. It's a case study in how federal efforts to enhance American competitiveness can go awry. [Mr. Pike, who is often quoted on space matters, failed to note that the Europeans and Japanese were conducting many elements of the ACTS program for their own industries. He must have thought that they were wasting a lot of their money also. In an ABC interview in 1993, Mr. Pike also said that the Ka-band was too 'fragile' and would not be used for many, many years. He failed to note that Iridium, which was being built at the time, used Ka-band feeder and cross-links.]
>
> Nevertheless, analysts say a moral of the ACTS story is that government should tread very gingerly when it tries to help industry technically and that any aid programs that do materialize should be structured so that businesses pay a substantial part of the costs, creating an

247

opening for the discipline of market mechanisms. Otherwise, they say, the white elephants are likely to multiply.

This view that the government is not capable of guiding technology for commercial markets led to a strong condemnation of the ACTS program by people who appeared to be philosophical zealots rather than qualified to judge the value of technology. By September of 1994, a little more than a year after this critical *New York Times* article, 15 separate filings were made to the FCC for Ka-band satellites very similar in architecture to ACTS. By 1998, there had been filings made to the ITU for more than 700 Ka-band satellites. It is recognized, of course, that not all these satellites will be built. These filings, however, are a strong indication that there is a future for Ka-band satellites. Although the government has made its share of mistakes (as has industry) in picking technology winners, it isn't always true that government programs are a bust. As demonstrated by ACTS, government and industry can join together for the betterment of all.

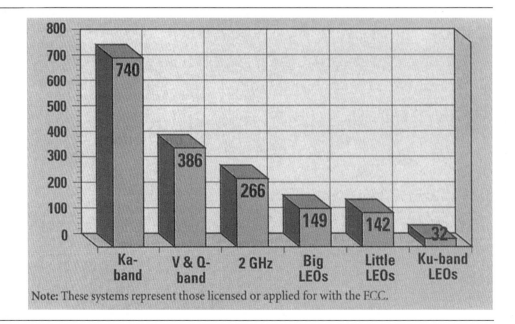

Number of Planned, Non-Traditional Satellites (source: *Via Satellite*, July 1998)

Note: These systems represent those licensed or applied for with the FCC.

Congress Champions ACTS Program

Congress decided the fate of ACTS. Congress was increasingly concerned about U.S. economic competitiveness in high technology industries in the 1980s because U.S. prosperity was increasingly dependent on those industries. Therefore, Congress was sensitive to areas such as satellite communication being challenged by foreign entities. The Democratic Congress listened to the arguments of U.S. private industry in support of the ACTS program and decided the ACTS program should be initiated. The congressional committees felt that the goals established by NASA, in concert with industry, were in the national interest and essential to maintaining U.S. preeminence in the communication satellite arena. Congress also believed a flight test program was required to give service providers the confidence to implement these technologies in future commercial systems.

Role of Government Technology Sponsorship in the Future

Today's world is dominated by intense global competition and fueled by technology changes. The satellite industry remains increasingly competitive. While satellite communication is now thought of as a "mature" industry, aggressive expansion of global satellite networks is making the satellite marketplace a dynamic, revolutionary, and risky environment. U.S. manufacturers continue to lead, but Japanese companies are expanding their presence, and European manufacturers are consolidating to enhance their ability to compete. Competition from landline and terrestrial wireless technology also remains high. Success depends on manufacturers producing cost-efficient systems, while maintaining high system reliability. The fast pace of today's market requires the continual renewal of the technology base in order for companies to keep pace with global growth and stay competitive.

President Clinton, discussing his U.S. commercial technology policy in 1993, explained that American technology must move in a new direction to build economic strength and spur economic growth [163]. He indicated that NASA must modify the way it does business and encourage cooperative work with industry in areas of mutual interest—a significant challenge to NASA. He went on to point out that a fundamental mechanism for carrying out this new approach at the level of technology applications is a cost-shared R&D partnership between government and industry.

Satellite Industry Task Force

In 1995, aerospace and communications executives from 21 communication satellite companies (and several related industry companies) recognized the need for a collective approach to addressing key industry issues and formed a Satellite Industry Task Force (SITF). In a 1996 presentation to Vice President Albert Gore, with DOD and NASA executives present, Dr. Thomas Brackey of Hughes Space & Communications (chairman of the SITF), acknowledged that a strong research role in the field of satellite communications was appropriate

249

for the U.S. government at a time when communication satellites were entering a key new stage of growth and expanding into new applications [164].

Satellite Industry Task Force (SITF) Members

1. American Mobile Satellite

2. AT&T Corporation

3. Ball Aerospace

4. Bellcore

5. Boeing Company

6. Computer Sciences Corporation

7. COMSAT Corporation

8. CTA Corporation

9. GE American Communications

10. Globalstar L.P.

11. Hughes Space & Communications

12. Hughes Network Systems

13. Lockheed Martin Corporation

14. Lucent Technologies

15. MCI Communications

16. Motorola Inc.

17. Orbital Sciences

18. Orion Satellite

19. Space Systems/Loral

20. Teledesic

21. TRW Inc.

NASA, along with DOD, seemed to accept the basic premises of the SITF briefing to Vice President Gore, especially in regard to greater industry/government interaction in the planning and conduct of satellite technology R&D. As part of its response, NASA initiated the concept for a technology development alliance, which was referred to as Satellite Alliance USA.

Government-Industry Alliance

The Alliance was to provide new options for leveraging resources with partners from industry, government, and academia in industry-led technology development projects. The Alliance sought to combine the talents and resources of its members toward achieving key goals identified by the SITF as necessary to sustaining U.S. leadership in satellite-based communications.

Accordingly, Satellite Alliance USA sought to:

- realize the full potential of satellite communications in the national and global information infrastructures

- enhance the global competitiveness of the U.S. satellite industry

- maintain a strong national research base through coordinated support for satellite and related technology innovation, development, and demonstrations

Through the Alliance, strategic national priorities for satellite technology were to be set, laying the groundwork for joint development of critical high-risk technologies. A key feature of the Alliance structure was that different government agencies, industry, and academia could combine forces by linking to the underlying Alliance agreement, which was to be initiated with a minimum of infrastructure and a maximum of flexibility for participants.

As part of the Alliance, one of the key programs that NASA was to embark on was to initiate precompetitive technology development for the next generation of space-based communication networks from the perspective of the global information infrastructure. These precompetitive technologies targeted for the ten year plus time frame were intended to reduce the technical risks and act as a catalyst to open up new markets for the U.S. satellite communications industry. The term *precompetitive technologies* is not well defined, but we believe it means the development of device- and assembly-level technologies as opposed to system-level flight spacecraft like ACTS. This reflects the current view that given the flight-proven component technologies, the industry is now in a position to assume system-level technical risks. There is no quarrel with this position, although system-level considerations need to be defined in order to identify the necessary component and subsystem technologies.

251

Although considerable time and work was performed by the government, industry, and academia, the formation of the Satellite Alliance was halted when promised NASA funds did not materialize and significant leadership changes occurred in both the DOD and OSTP. In 1998, NASA essentially made the decision to fund only technology that would directly benefit specific NASA missions. This decision appears to end the ACTS era of direct government support for high-risk, long-term R&D of satellite communication technology, which specifically benefits the commercial sector. The NASA decision was probably influenced by budget priorities with its own mission-specific programs—such as the space station—being of higher importance. The DOD decision to withhold funding and leadership from the Satellite Alliance was influenced by the newly initiated Space Technology Alliance (STA).

As both government and industry were abandoning the Satellite Alliance, the DOD was forming a broader based, government-led initiative called the Space Technology Alliance (STA). Under the leadership of Christene Anderson of the USAF, Phillips Laboratory, the STA is comprised of leaders from government agencies and the DOD, who have both a vested interest in space and a substantial related technology program. The stated purpose of the STA is to identify and coordinate the related technology programs, including classified ones, of the various government agencies to maximize the benefit of the government's technology investments for all of its missions. In contrast to the Satellite Alliance, the STA will examine a broad array of space-related technologies—many relevant to the communications satellite industry, such as power and onboard propulsion—and others of lesser interest, such as hyperspectral imaging. Therefore, communication technologies planned for the Satellite Alliance may now be developed by the STA. Once the technology needs for the government are determined, the STA plans to use panels of aerospace industry experts (including some SITF members) to review the selected technology areas. Industry's contribution will include providing the STA with an economic and market perspective of the government's technology plans and funding investments. The STA encourages, but does not require consolidation of resources. Each government agency is free to pursue its needs independently without industry involvement or sharing of resources. However, the declining federal budget and its resultant effect on government and industry technology programs is causing a strong motivation for the government to increase interdependency and leveraging of resources for the benefit of both government and industry.

Continuing Debate

The robustness of the U.S. economy is due, in large part, to its technological capabilities, but even in the era of global companies it will be necessary for government to fund long-term R&D if that U.S. robustness is to be maintained. And the debate will go on as to the proper involvement of government, and it should. In 1999 the President's Information Technology Advisory Council, made up of high-tech business executives, university professors, and researchers has called for increased funding [165]. Its findings, which present arguments similar to the ones made for ACTS, are summarized as follows:

> 'In both the public and private sectors today, U.S. investments in technology R&D have slowed to a relative trickle,' the report says. The end of the Cold War signaled a demise in research funding. From 1987 to 1996, federal research funding fell 2.6 percent a year.
>
> If the trend away from long-term research continues, the flow of bold ideas that has fueled the new economy will slow to a trickle by 2010, which would risk losing the nation's competitive edge to other countries.
>
> The industry cannot fund long-term research because its economic reality dictates a short-term focus. The federal government must expand its role in leading long-term research. Federal investment directly supports the education and training of people to enter the information technology (IT) marketplace.
>
> 'Trained people are not just a by-product, but rather a major product of publicly supported research. These trained professionals are critical national infrastructure, and will create and develop new ideas, form a talent pool for existing business, and launch new companies,' the council said.

As far as the government's funding of advanced space technology, the United States has currently decided to spend its funds on mission-specific projects that meet the needs of the government. In the case of NASA this is driven by budget considerations. Whether developing technology chiefly to meet government needs is a wise decision will mostly depend on how much mission-specific, government technology spins-off for commercial applications. If the industry feels seriously threatened by foreign competition, there is no doubt that political forces will increase the government's funding for satellite R&D.

The particular economic, political, and competitive forces and technological challenges dictate the correct policy at any particular time. Currently we are in an era where the forces are tending to minimize government involvement. However, as ATS, CTS, and ACTS have demonstrated, long-term, high-risk government R&D has an invaluable benefit to the nation. No one, in this time of minimizing the role of government, should forget the proven positive benefits of long-term government R&D to the nation's economy.

253

APPENDIX A

SPACE TECHNOLOGY HALL OF FAME INDUCTEES

Sponsored by:
United States Space Foundation
2860 South Circle Drive
Colorado Springs, CO 80906

Since 1988, the Space Technology Hall of Fame has honored individuals, organizations, and companies responsible for remarkable products developed from space technology.

Space spinoffs are materials and products originally developed for space program requirements that have made significant contributions to humankind through commercial application.

Each year, such space technologies are nominated for induction into the Space Technology Hall of Fame. A selection committee of nationally prominent citizens reviews the nominations against stringent criteria and identifies the technologies to be inducted. The technology innovators are recognized for creative design and business applications. In 1997, ACTS was inducted into the Space Technology Hall of Fame. The inductees were:

NASA Jet Propulsion Laboratory
NASA Lewis Research Center, Donald J. Campbell, Director
Bolt Baranek Neuman
COMSAT Laboratories, Benjamin Pontano, President
EMS Technologies, Thomas E. Sharon, Ph.D., CEO
Harris Corporations, Donald C. Schultz
Lockheed Martin Missiles and Space, Michael Henshaw, President
Mayo Clinic
Motorola Space & Systems Technology Group, Durrell W. Hillis, Senior Vice President & General Manager

Brian Scott Abbe

Roberto J. Acosta, Ph.D.

Martin J. Agan

Stanley W. Attwood

Robert A. Bauer

George A. Beck

David R. Beering

Michael J. Berrett

Ronald Lee Bexten

Marcos Antonio Bergamo, Ph.D.

Larry C. Brown

Daniel L. Brandel

William A. Bucher

Philip Carvey

Thom A. Coney

James Glen Cory

Ramon Perez De Paula, Ph.D.

Khaled Dessouky, Ph.D.

Irving Dostis

Burton I. Edelson, Ph.D.

Jack Frohbieter

Joshua B. Gahm

Frank Gargione

Richard T. Gedney, Ph.D.

James R. Greaves

William Hawersaat

Robert E. Hay

Durrell W. Hillis

Douglas J. Hoder

Dave Huff

Kenneth Brian Ilse

Howard O. Jackson

Michael A. Jarrell

Thomas C. Jedrey

Russell J. Jirberg

Michael E. Kavka

Denis Kermicle

Bijoy K. Khandheria, Ph.D.

Rodney M. Knight

Richard J. Krawczyk

Hans Kruse, Ph.D.

Robert S. Lawton

Kerry D. Lee

Richard David Lilley

Richard R. Lindstrom

Robert Roland Lovell

Kevin M. McPherson

David Neal Meadows

Richard Lee Moat

Fiirouz Michael Naderi

Dean Allen Olmstead

Mark S. Plecity

Joanne R. Poe, J.D.

Charles E. Profera, Ph.D.

Karl Frederick Reader

Richard C. Reinhart

Ira Richer, Ph.D.

Dennis Dale Robinson

Bruce R. Savage

Ralph P. Schaefer

Ronald J. Schertler

Charles A. Schmidt

William G. Schmidt

James T. Shaneyfelt

Thomas E. Sharon, Ph.D.

Joe Sivo

Jon Michael Smith

Philip Y. Sohn

Ernie Spisz

Steven Storch

David L. Wright

Michael Zernic

APPENDIX B

GLOBAL GEOSTATIONARY ORBIT KA-BAND FILINGS AS OF MID-1997 [151]

SATELLITE SYSTEM	COUNTRY of ORIGIN	SPONSORS	ORBITAL SLOTS REQUESTED
AFRISAT	UNITED KINGDOM (HK)		4
ARABSAT Ka	ARAB LEAGUE	ARABSAT	12
ASIASAT-AKA	UNITED KINGDOM (HK)		5
ASTROLINK	USA	LOCKHEED MARTIN	5
BIFROST	NORWAY	TELENOR	1
CANSAT Ka	CANADA	TELESAT CANADA	5
CHINASAT 41-47	CHINA	CHINESE GOVT.	7
COMETS	JAPAN	NASDA	1
CYBERSTAR	USA	LORAL SPACE	3
DACOMSAT	SOUTH KOREA	DACOM	1
DB-SAT	GERMANY		1
DRTS	JAPAN		5
EAST	UNITED KINGDOM		5
EASTSAT	KOREA		1
ECHOSTAR-Ka	USA	ECHOSTAR	2

SATELLITE SYSTEM	COUNTRY of ORIGIN	SPONSORS	ORBITAL SLOTS REQUESTED
EDRSS	ESA	ESA	4
ETS-8	JAPAN	NASDA	3
EUROSKYWAY	ITALY	ALENIA SPAZIO	7
EUROPESTAR K	GERMANY		3
EUTELSAT-Ka	REGIONAL, EUROPE	EUTELSAT	22
GENESIS	GERMANY	DT	1
GE STAR	USA	GE AMERICOM	5
GLOBALSAT	KOREA		1
HISPASAT-2AKa	SPAIN	HISPASAT	1
INFOSAT	KOREA		3
INMARSAT-GSO-2	INTERNATIONAL	PTT'S	5
INSAT, INSAT-Ka	INDIA	INSAT/ISRO	15
INTELSAT-Ka	INTERNATIONAL	PTT'S	12
JBS	USA		17
KaSATCOM	USA		6
KaSTAR (New name iSKY)	USA	KaSTAR	2
KOREASAT	SOUTH KOREA	KOREASAT	1
KYPROS-SAT-Ka	CYPRUS		5
KYPROS-SAT-L	CYPRUS		4
LUX-Ka(ASTRA)	LUXEMBOURG	SES	20
MALTASAT-1	MALTA		4

SATELLITE SYSTEM	COUNTRY of ORIGIN	SPONSORS	ORBITAL SLOTS REQUESTED
MEASAT	MALAYSIA	SHINAWATRA	5
MEASATLA&SA	MALAYSIA	SHINAWATRA	5
MEDSAT	MEDITERRANEAN	AEROSPATIALE, etc.	1
MEGASAT	MEXICO		10
MILLENNIUM	USA	MOTOROLA	4
MORNING STAR	USA		4
MTSAT	JAPAN		3
NETSTAR-28	USA		1
ORION	USA	ORION	6
PAKSAT	PAKISTAN		5
PANAMSAT	USA	PANAMSAT(HUGHES)	2
ROSCOM	RUSSIA		4
SAMSAT	SINGAPORE	PACIFIC CENTURY	3
SARIT	ITALY	TELESPAZIO	1
SICRAL-2	ITALY		2
SIRIUS-4	SWEDEN	NSAB	1
SKYSAT	UNITED KINDOM	AFRO-ASIAN	12
SPACEWAY	USA	HUGHES	14
STENTOR	FRANCE		1
SUPERBIRD	JAPAN	SCC	2
SYRACUSE	FRANCE		9
THAICOM Ka	THAILAND	BINARIANG	5

SATELLITE SYSTEM	COUNTRY of ORIGIN	SPONSORS	ORBITAL SLOTS REQUESTED
TONGASAT Ka-1	TONGA	TONGASAT	9
TOR-M	RUSSIA		18
TURKSAT-Ka	TURKEY	TURKSAT	6
USCSID	USA		12
USGBS	USA		5
USGCB	USA		5
VIDEOSAT Ka	FRANCE	FRANCE TELECOM	3
VINASAT-A	REGIONAL	VIETNAM	4
VISIONSTAR	USA		1
WEST-GEO	FRANCE	MATRA MARCONI	12
YAMAL	RUSSIA	GAZPROM	5

ACRONYM LIST

ACTS	Advanced Communications Technology Satellite
AGC	Automatic Gain Control
ANL	Automatic Number Location
AIAA	American Institute of Aeronautics and Astronautics
AKM	Apogee Kick Motor
AM	Amplitude Modulation
AMT	ACTS Mobile Terminal
APT	ACTS Propagation Terminal
APU	Auxiliary Power Unit
APW	ACTS Propagation Workshop
ARIES	ATM Research and Industrial Enterprise Study
ARPANET	Advanced Research Project Agency's Network
ASIC	Application Specific Integrated Circuit
ATM	Asynchronous Transfer Mode
ATS	Application Technology Satellite
AWGN	Additive White Gaussian Noise
B	Billion
BBP	Base Band Processor
BER	Bit Error Rate
BFN	Beam Forming Network
B-ISDN	Broadband Integrated Services Digital Network
BOL	Beginning of Life
Bps	Bits Per Second
BPSK	Binary Phase-Shift Keying
BSS	Broadcast Satellite Service
BTC	Budget to Completion
CATN	Common Antenna Television Network
CDR	Critical Design Review
CD-ROM	Compact Disk-Read Only Memory
CML	Current Mode Logic
CMDS	Commands
CMOS	Complementary Metal Oxide Semiconductor
CODEC	Coder-Decoder
CONUS	Continental United States

CPS	Customer-Premises Services
CPU	Central Processing Unit
CRT	Cathode Ray Tube
CR&T	Command Ranging & Telemetry
CT	Computed Tomograghy
CTS	Communication Technology Satellite
CW	Continuous Wave
CWG	Carrier Working Group
CY	Calendar Year
DAMA	Demand Assigned Multiple Access
DARPA	Defense Advanced Research Projects Agency
dB	Decibel
dBi	Decibel, Isotropic
DBS	Direct Broadcast Service
dBW	Decibel, Watt
DC	Direct Current
D/C	Downconverter
DCU	Digital Control Unit
DCTN	Defense Communication Telecommunications Network
DOD	Department of Defense
DPSK	Differential Phase-Shift Keying
DTH	Direct to Home
Eb/No	Energy per bit per Noise Density
ECL	Emitter Coupled Logic
EDT	Eastern Daylight Time
EIA	Electronic Industries Association
EIRP	Effective Isotropic Radiated Power
EOA	Experiment Opportunity Announcement
ESA	Earth Sensor Assembly, European Space Agency
FAX	Facsimile
FCC	Federal Communications Commission
FCS	Fixed Communication Service
FDM	Frequency Division Multiplexing
FDMA	Frequency Division Multiple Access
FDR	Final Design Review
FEC	Forward Error Correction
FET	Field Effect Transistor
G/T	Antenna Gain/Receiving System Noise Temperature
GaAs	Gallium Arsenide
GEO	Geostationary Earth Orbit
GES	Gigabit Earth Station

GHz	Gigahertz
GPS	Global Positioning System
GSFC	NASA Goddard Space Flight Center
GSO	Geostationary Orbit
GTO	Geostationary Transfer Orbit
GII	Global Information Infrastructure
HDR	High Data Rate
HDV	High Definition Video
HEMT	High Electron Mobility Transistor
HMMWV	Highly Mobile Maneuverable Wheeled Vehicle
HV EPC	High Voltage Electronic Power Conditioner
Hz	Hertz (cycles per second)
I&T	Integration and Test
IBOW	In Bound Order Wire
IF	Intermediate Frequency
IMPATT	Impact Avalanche Transit Time (Device)
INS	Inertial Navigation System
IP	Internet Protocol
IR&D	Internal Research and Development
IRP	Initial Response to Proposal
ISDN	Integrated Services Digital Network
IT	Information Technology
ITU	International Telecommunication Union
JPL	Jet Propulsion Laboratory
JSC	NASA Johnson Space Center
JWID	Joint Warrior Interoperability Demonstration
KAO	Kuiper Airborne Observatory
Kbps	Kilobits Per Second
KHz	Kilohertz
LAN	Local Area Network
LeRC	NASA Lewis Research Center
LEO	Low Earth Orbit
LET	Link Evaluation Terminal
LLNL	Lawrence Livermore National Laboratory
LNA	Low Noise Amplifier
LNR	Low Noise Receiver
LRD	Launch Readiness Date
LSI	Large Scale Integration
mm	Millimeters
ms	Millisecond
M	Million

M/V	Motor Vessel
MAN	Metropolitan Area Network
MBA	Multiple Beam Antenna
Mbps	Megabits Per Second
MCS	Master Control Station
MHz	Megahertz
MMIC	Monolithic Microwave Integrated Circuit
MOSAIC	Motorola Oxide Self-Aligned Implanted Circuit
MR	Magnetic Resonance
MRI	Magnetic Resonance Imaging
MSM	Microwave Switch Matrix
MSP	Modular Switching Peripheral
MSS	Mobile Satellite System
MTS	Metered Telephone Service
ns	Nanosecond
NAPEX	NASA Propagation Experiment
NASA	National Aeronautics & Space Administration
NASDA	National Space Development Agency (Japan)
NCAR	National Center for Atmospheric Research
NCS	National Communication Systems
NGS	NASA Ground Station
NGSO	Non Geostationary Orbit
NII	National Information Infrastructure
NLH	Net Long Haul
NMT	Network Management Terminal
NOI	Notice of Intent
NRC	National Research Council
OBOW	Out Bound Order Wire
OC	Optical Character
OMB	Office of Management and Budget
PAM	Payload Assist Module
PBS	Public Broadcast System
PC	Personal Computer
PIN	Personal Identification Numbers
PD	Presidential Directive
PDR	Preliminary Design Review
PHEMT	Pseudomorphic High Electron Mobility Transistor
PM	Phase Modulation
POC	Proof of Concept
PSN	Public Switched Network
QPSK	Quaternary Phase-Shift Keying

rms	Root-Mean-Square
R&D	Research and Development
RCSA	Receive Coax Switch Assembly
RF	Radio Frequency
RFT	Radio Frequency Terminal
RFP	Request for Proposals
SBA	Steerable Beam Antenna
SBPSK	Staggered Binary Phase-Shift Keying
SBS	Satellite Business System
SCADA	Supervisory Control and Data Acquisition
SDR	System Design Review
SES	Societe Europeenne des Satellites
SITF	Satellite Industry Task Force
SMSK	Serial Minimum Shift Keying
SNMP	Simple Network Management Protocol
SONET	Synchronous Optical Network
SQPSK	Staggered Quaternary Phase-Shift Keying
SSI	Small Scale Integration
SS-TDMA	Satellite Switched Time Division Multiple Access
STS	Space Transportation System
STU	Secure Telephone Unit
T1	1.544 Mbps Bell Standard Digital Data Rate
TCP	Transmission Control Protocol
TCSA	Transmit Coax Switch Assembly
TDM	Time Division Multiplexing
TDMA	Time Division Multiple Access
TDRSS	Tracking & Data Relay Satellite System
TEW	Tracking Error Word
TIP	Telemedicine Instrument Package
TLM	Telemetry
TOS	Transfer Orbit Stage
TT&C	Telemetry, Tracking & Command
TWT	Traveling Wave Tube
TWTA	Traveling Wave Tube Amplifier
TV	Television
TWX	Teletypewriter Exchange
U/C	Upconverter
UDP	User Datagram Protocol
UHF	Ultra High Frequency
USAT	Ultra Small Aperture Terminal
VHF	Very High Frequency

VPN	Virtual Private Network
VSAT	Very Small Aperture Terminal
WATS	Wide Area Telephone System
WIRS	Waveguide Input Redundancy Switch
WORS	Waveguide Output Redundancy Switch
Z	Zulu or Greenwich Mean time
3D	Three Dimensions
°K	Degrees Kelvin
µs	Microsecond

REFERENCES

1. Linda R. Cohen and Roger G. Noll, *The Technology Pork Barrel* (Washington, D.C.: The Brookings Institution, 1991).

2. Armand Musey, *The Satellite Book* (C.E. Unterberg, Towbin Research Company, 1999).

3. David Martin, *Communication Satellites 1958–1992* (El Segundo CA: The Aerospace Corporation, 1991).

4. Guy Beakley, "Satellite Communications, Growth and Future," *Telecommunications* (November 1980).

5. Peter Cunniffe, "Misreading History: Government Intervention in the Development of Commercial Communication Satellites," *Program in Science and Technology for International Security* (Massachusetts Institute of Technology) 24, (May 1991).

6. Philip Chien, "The History of Geostationary Satellites," *Launch Space Magazine* (February/March 1998).

7. Philip Chien, "Syncom at 30: Some Milestones in Satellite History," via satellite. (July 1993).

8. NASA Lewis Research Center, *CTS Reference Book*, NASA Technical Memorandum X-71824 (Cleveland, Ohio: October 15, 1975).

9. Robert Lovell and C. Louis Cuccia, "Type-C Communication Satellites Being Developed for the Future," *Microwave System News* (March 1984).

10. Chris Bulloch, "Advancing the Art of Satellite Communications—Foreign Competition Spurs NASA SATCOM Research," *Interavia* 40 (January 1985).

11. Daniel Branscome, "The Evolving Role of the Federal Government in Space Communications Research and Development, *Proceedings of the 28th American Astronautical Society Annual Conference*, San Diego, California (October 26-29, 1981).

12. Committee on Satellite Communications of the Space Application Board, "Federal Research and Development for Satellite Communications," Report, *Assembly of Engineering, National Research Council* (1977).

13. "NASA Space Communications R&D: Issues, Derived Benefits and Future Directions," Space Applications Board, Commission

on Engineering and Technical Systems of the National Research Council, National Academy Press (February 1989).

14. Joseph Sivo, "Advanced Communications Satellites," *NASA Technical Memorandum 81599* (1980).

15. Joseph Sivo, "30/20 GHz Experimental Communication Satellite System," *Technical Memorandum 82683* (1981).

16. T. Gabriszeski, P. Reiner, J. Rogers, and W. Terbo, "18/30 GHz Fixed Communications System Service Demand Assessment," Western Union Telegraph Company, *NASA Contract Report 159546* (1979).

17. Robert Gamble, H. Seltzer, K. Speter, and M. Westheimer, "18/30 GHz Fixed Communications System Service Demand Assessment," U.S. Telephone and Telegraph Corporation/ITT, *NASA Contract Report 159620* (1979).

18. "18/30 GHz Fixed Satellite Communications System Study," Hughes Aircraft Company, *NASA Contract Report 159672* (1979).

19. "18/30 GHz Fixed Satellite Communications System Study," Ford Aerospace and Communications Corporation, *NASA Contract Report 159625* (1979).

20. Dominick Santarpia and James Bagwell, "NASA's Multibeam Communications Technology Program," *Microwave Journal* (January 1984).

21. James Bagwell, "Technology Achievements and Projections of Communication Satellites of the Future," *Paper 86-0649*, 11th AIAA International Communication Satellite Systems Conference (1986).

22. Robert E. Berry, Director of Space Systems Operations at Ford Aerospace & Communications Corporation. Letter to Robert L. Firestone, contract specialist at NASA Lewis Research Center (May 3, 1983).

23. Jay C. Lowndes, "NASA Plans Test SATCOM Award," *Aviation Week & Space Technology* (December 12, 1983): 17.

24. David Reudink, A. Acampora, and Y. Yeh, "The Transmission Capacity of Multibeam Communication Satellites," *Proceedings of the IEEE* 69, no. 2 (February 1981).

25. David Wright and Joseph Bolombin, "ACTS System Capabilities and Performance," *Proceedings of the 14th AIAA International Communications Satellite System Conference*, Washington, D.C. (March 1992).

26. F. Michael Naderi and S. Joseph Campanella, "NASA's Advanced Communications Technology Satellite (ACTS): An Overview of the Satellite, the Network, and the Underlying Technologies," *Proceedings*

of the 12th AIAA International Communications Satellite System Conference, Arlington, Virginia (March 1988).

27.   John Grabner and William Cashman, "ACTS Multibeam Communications Package: Technology for the 1990s," *Proceedings of the 13th AIAA International Communications Satellite Systems Conference*, Los Angeles, California (March 1990).

28.   Peter Lowry, *ACTS System Handbook* (Cleveland, Ohio: NASA TM-101490, NASA LeRC, 1991).

29.   Frank Regier, "The ACTS Multibeam Antenna," *NASA Technical Memorandum 105645* (April 1992).

30.   Charles Profera and Sutinder Dhanjal, "ACTS Multibeam Antenna Subsystem," *Proceedings of the NASA LeRC ACTS Results Conference*, Cleveland, Ohio (September 1995).

31.   Guy Blatt and Jack Harper, "Multibeam Antenna Reflectors," *Proceedings of the NASA LeRC ACTS Results Conference*, Cleveland, Ohio (September 1995).

32.   Paul Cox and Kurt Zimmerman, "ACTS Scanning Beam-Forming Network Technology Update," *Proceedings of the NASA LeRC ACTS Results Conference*, Cleveland, Ohio (September 1995).

33.   Youn Choung, Herschel Stiles, Joseph Wu, William Wong, C. Harry Chen, and Ken Oye, "NASA ACTS Multibeam Antenna (MBA) System," *Proceedings of the IEEE Eastcom-86*, Washington, D.C. (September 1986).

34.   Alan Hewston, Kent Mitchel, and J. Sawicki, "Attitude Control Subsystem for the Advanced Communications Technology Satellite," *NASA TM 107352* (November 1996).

35.   Richard Moat, "ACTS Baseband Processor," *Proceedings of the IEEE Global Telecommunications Conference* vol. 1 (December 1986) 578-583.

36.   Thomas Inukai, David Jupin, R. Lindstrome, and David Meadows, "ACTS TDMA Network Control Architecture," *Proceedings of the 12th AIAA International Communications Satellite Systems Conference*, Arlington, Virginia (March 1988).

37.   Kerry Lee, "Design and Development of a Baseband Processor for the Advanced Communications Technology Satellite," *Proceedings of the NASA LeRC ACTS Results Conference*, Cleveland, Ohio (September 1995).

38.   Michael Moreken and John Zygmaniak, "ACTS IF Switch Matrix," *Proceedings of the NASA LeRC ACTS Results Conference*, Cleveland, Ohio (September 1995).

269

39.   Richard Gedney, "Results from ACTS Development and On-Orbit Operations," *Proceedings of the 1st Ka-Band Utilization Conference*, Rome, Italy (November 1995).

40.   Frank Gargione, Roberto Acosta, Thom Coney, and Richard Krawczyk, "Advanced Communication Technology Satellite (ACTS): Design and On-Orbit Performance Measurements," *International Journal of Satellite Communications* 14 (1996) 133-159.

41.   Richard Krawczyk, "ACTS Operational Performance Review," *Proceedings of the NASA LeRC ACTS Results Conference*, Cleveland, Ohio (September 1995).

42.   Richard Krawczyk and Frank Gargione, "Performance of ACTS as a Ka-Band Test Bed," *Proceedings of the 5th Ka-Band Utilization Conference*, Taormina, Italy, (October 18-20, 1999).

43.   Richard Gedney and Thom Coney, "Effective Use of Rain Fade Compensation for Ka-Band Satellites," *Proceedings of the 3rd Ka-Band Utilization Conference*, Sorrento, Italy (September 1997).

44.   Chrisina Cox and Thom Coney, "Advanced Communications Technology Satellite Adaptive Rain Fade Compensation Protocol Performance," *Proceedings from 4th Ka-Band Utilization Conference*, Venice, Italy (November 2-4, 1998).

45.   Thom Coney, "Advanced Communications Technology Satellite (ACTS) Very Small Aperture Terminal (VSAT) Network Control Performance, *Proceedings of the 16th AIAA International Communications Satellite System Conference*, Washington, D.C. (March, 1996).

46.   Doug Hoder and Marcus Bergamo, "Gigabit Satellite Network for NASA's Advanced Communications Technology Satellite (ACTS)," *International Journal of Satellite Communications* 14 (1996) 161-173.

47.   Roberto Acosta, "Advanced Communications Technology Satellite (ACTS) Multibeam Antenna Analysis and On-Orbit Performance," *Proceedings of the 2nd Ka-Band Utilization Conference*, Florence, Italy (September 1996).

48.   Walter L. Morgan and Denis Rouffet, *Business Earth Stations for Telecommunications*, (New York: John Wiley & Sons, 1988).

49.   S. O'Rourke and David Hartshorn, "Future Satellite Applications: Is the Writing on the Wall," *Satellite Communications* (August 1999).

50.   Richard D. Lilley and Dennis D. Robinson, "Design Consideration on the ACTS T1-VSAT," *Proceedings of the NASA LeRC ACTS Results Conference*, Cleveland, Ohio (September 11-13, 1995).

51. R. Schaefer, R. Cobb, and Dennis Kermicle, "Link Quality Estimation for ACTS T1-VSAT," *Proceedings of the NASA LeRC ACTS Conference*, Washington, D.C. (November 18-19, 1992).

52. Philip P. Carvey, "Digital Terminal Architecture and Implementation in the ACTS High Data Rate Earth Station for the DARPA/NASA Gigabit Satellite Network," *Proceedings of the NASA LeRC ACTS Results Conference*, Cleveland, Ohio (September 11-13, 1995).

53. Brian D. May, "The Link Evaluation Terminal for the Advanced Communications Technology Satellite Experiments Program," *Proceedings of the 14th AIAA International Communications Satellite Systems Conference* 3, Washington, D.C. (March 22-26, 1992) 1749-1757.

54. Richard C. Reinhart, "System Design and Application of the Ultra Small Aperture Terminal with the ACTS," *Proceedings of the 3rd Ka-Band Utilization Conference*, Sorrento, Italy (September 1997).

55. Brian S. Abbe, Martin J. Agan, and Thomas C. Jedrey, "ACTS Mobile Terminals," *International Journal of Satellite Communications* 14 (1996) 175-189.

56. Deborah S. Pinck and Michael Rice, "Mobile Propagation Results Using the ACTS Mobile Terminal," *Proceedings of the NASA LeRC ACTS Results Conference*, Cleveland, Ohio (September 11-13, 1995).

57. Julius Goldhirsh and Wolfhard J. Vogel, "Mobile Propagation Measurements and Modeling Results Using ACTS," *Proceedings of the 16th AIAA International Communications Satellite Systems Conference* 1, Washington, D.C. (February 25-29, 1996) 379-387.

58. S. Sanzgiri, D. Bostrum, W. Pottenger, and R. Lee, "A Hybrid Tile Approach for Ka-Band Sub-array Modules," *IEEE Transactions on Antennas and Propagation* 43, no. 9 (September 1995).

59. Charles Raquet, Robert Zakrajsek, Richard Lee, Monty Andro, and John Turtle, "Ka-Band MMIC Array System for ACTS Aeronautical Terminal Experiment," *Proceedings of International Mobile Satellite Conference – IMSC'95*, Ottawa, Canada (June 6-8, 1995) 312-317.

60. David N. Meadows, "The ACTS NASA Ground Station/Master Control Station," *Proceedings of the 14th AIAA International Communications Satellite Systems Conference* 3, Washington, D.C. (March 22-26, 1992). 1172-1182.

61. Steven. J. Struharik and David N. Meadows, "The ACTS Master Ground Station – Review of System Design and Operational Performance," *Proceedings of the 1st Ka-Band Utilization Conference*, Rome, Italy (October 10-12, 1995).

271

62. Robert Bauer, "Ka-Band Propagation Measurements: An Opportunity with the Advanced Communications Technology Satellite (ACTS)," *Proceedings of the IEEE* 85, no. 6 (June 1997).

63. Robert K. Crane, Suhe Wang, Davis B. Westenhaver, and Wolfhard J. Vogel, "ACTS Propagation Experiment: Experiment Design, Calibration, and Data Preparation and Archival," *Proceedings of the IEEE* 85, no. 6 (June 1997).

64. David B. Westenhaver, "ACTS Propagation Terminal," *Proceedings of the NASA LeRC ACTS Results Conference*, Cleveland, Ohio (September 11-13, 1995).

65. Ronald Schertler, "ACTS Experiments Program," *Proceedings of the NASA LeRC ACTS Results Conference*, Cleveland, Ohio (September 11-13, 1995).

66. Bijoy Khandheria, Marvin Mitchel, Abdul Bengali, Joseph Duffy, Thomas Kottke, Michael Wood, and Barry Gilbert, "Telemedicine: The Mayo Clinic Experience with Low Data Transmissions on the NASA ACTS Satellite," *Proceedings of the NASA LeRC ACTS Results Conference*, Cleveland, Ohio (September 11-13, 1995).

67. Thomas Kottke, Leonard Finger, Mary Trapp, Laurel Panser, and Paul Novotny, "The Pine Ridge-Mayo National Aeronautical and Space Administration Telemedicine Project: Program Activities and Participation Reaction," *Mayo Clinic Proceedings* 71, no. 4 (April 1996) 329-337.

68. Robert Kerczewski, Gerald Chomos, Paul Mallasch, Duc Ngo, Diepchi Tran, Quang Tran, and Brian Kachmar, "Operational Scenarios and Implementation Issues for T1-Rate Satellite Telemammography," *Proceedings of the 17th International Communications Satellite Systems Conference*, Yokohama, Japan (March 1998).

69. Bruce Jackson, "Emergency Medical Services," *Satellite Communications Magazine* (December 1994) 27-28.

70. David Yun and Hung Chen-Garcia, "Sharing Computational Resources and Medical Images Via ACTS-Linked Networks," *Proceedings of the Pacific Telecommunications Conference*, Honolulu, Hawaii (January 1996).

71. Kenneth Ilse, "The Army's Use of the Advanced Communications Technology Satellite," *Proceedings of the NASA LeRC ACTS Results Conference*, Cleveland, Ohio (September 11-13, 1995).

72. Brian Abbe and Deborah Pinck, "AMT Experiment Results," *Proceedings of the 4th International Mobile Satellite Conference*, Ottawa, Canada (June 6-8, 1995).

73. Charles Raquet, Konstantinos Martzaklis, Robert Zakrajsek, Monty Andro, and John Turtle, "MMIC Phased Array Demonstrations with ACTS," *Proceedings of the NASA LeRC ACTS Results Conference*, Cleveland, Ohio (September 11-13, 1995).

74. Christopher Passqualino, Brian Abbe, and Frank Dixon, "National Security/Emergency Preparedness and Disaster Recovery Communications Via ACTS," *International Journal of Satellite Communications* 14, no. 3 (May-June 1996) 219-232.

75. Hans Kruse, Tony Mele, Sara Young, Chris Washburn, and Donald Flournoy, "Disaster Recovery Applications for Satellite Communications Systems," *Proceedings of the NASA LeRC ACTS Results Conference*, Cleveland, Ohio. (September 11-13, 1995).

76. Stephen Horan, Michael Denny, Kurt Anderson, and James Fowler, "Real-Time Control of Remote Sites: Using the ACTS with the Apache Point Observatory," *Proceedings of the NASA LeRC ACTS Results Conference*, Cleveland, Ohio (September 11-13, 1995).

77. Christopher Carlin, Linda Hedges, and Issac Lopez, "Remote Propulsion Simulation," *Proceedings of the NASA LeRC ACTS Results Conference*, Cleveland, Ohio (September 11-13, 1995).

78. Larry Bergman, J. Patrick Gary, Burt Edelsen, Neil Helm, Judith Cohen, Patrick Shopbell, C. Roberto Mechoso, Choung-Chin, M. Farrara, and Joseph Spahr, "High Bit Rate Experiments Over ACTS," *International Journal of Satellite Communications* 14, no. 3 (May-June 1996) 259-266.

79. Eddie Hsu, Charles Wang, Larry Bergman, Naoto Kadowaki, Iida Takahashi, Burton Edelsen, Neil Helm, J. Pearman, and Frank Gargione, "Distributed HDV Post-Production Over Trans-Pacific ATM Satellites," *Proceedings of the 3rd Ka-band Utilization Conference*, Sorrento, Italy (September 1997).

80. Wayne Carlson, Stephen Spencer, Margaret Geroch, Matthew Lewis, Keith Bedford, David Welsh, John Kelly, and Arun Welsh, "Visualization of Results from a Distributed, Coupled, Supercomputer-Based Mesoscale Atmospheric and Lake Models Using the NASA ACTS," *Proceedings of the NASA LeRC ACTS Results Conference*, Cleveland, Ohio (September 11-13, 1995).

81. David Beering, "Satellite/Terrestrial Networks for Oil Exploration," *Proceedings of the NASA LeRC ACTS Results Conference*, Cleveland, Ohio (September 11-13, 1995).

82. David Beering, "Amoco Builds an ATM Pipeline," *Data Communications Magazine* (April 1995) 112-120.

273

83. "A New Way to Search for Oil, Via Satellite," *New York Times, Business Day,* (April 8,1996).

84. Charles Raquet, Robert Zakrajsek, Richard Lee, Monty Andro, and John Turtle, "Ka-Band MMIC Array System for ACTS Aeronautical Terminal Experiment (Aero-X)," *Proceedings of the 4th International Mobile Satellite Conference*, Ottawa, Canada (June 6-8, 1995).

85. Martin Agan, Daniel Nakamura, Alan Campbell, Robert Sternowski, Wendy Whiting, and Leon Shameson, "ACTS Aeronautical Experiments," *International Journal of Satellite Communications* 14, no. 3 (May-June 1996) 233-247.

86. R. Axford, G. Evanoff, R. North, K. Owens, J. Toy, G. Bostrom, T. Englun, W. Schmalgemeier, B. Hopkins, J. Griffin, M. Kelly, and P. Moose, "K/Ka-Band Maritime Mobile Satellite Communications," *Proceedings of the 5th International Mobile Satellite Conference,* Pasadena, California (June 6-8, 1995).

87. Roosevelt Fernandes, D. Brundaage, Barry Fairbanks, and Ronald Schertler, "Southern California Edison/NASA ACTS Experiment Low Cost SCADA Network," *Proceedings of the NASA LeRC ACTS Results Conference*, Cleveland, Ohio (September 11-13, 1995).

88. Harold Bradley SJ and Amy Kaufman, "Use of the Advanced Communications Technology Satellite to Promote International Distance Education Programs for Georgetown University," *Proceedings of the ACTS Results Conference*, Cleveland, Ohio (September 11-13, 1995).

89. Marc Chernick and Kevin Kenney, "Latin America Satellite Education Project," *A Final Report to NASA*, Georgetown University (October 1997).

90. Kul Bhasin, Daniel Glover, William Ivancic, and Tom vonDeak, "Enhancing End-to-End Performance Services Over Ka-Band Global Satellite Networks," *Proceedings of the 3rd Ka-band Utilization Conference*, Sorrento, Italy (September 1997).

91. David Brooks, T. Carrozzi, F. Dowd, Issac Lopez, S. Pelligrino, and Saragur Srinidhi, "ATM Based Geographically Distributed Computing Over ACTS," *Proceedings of the NASA LeRC ACTS Results Conference*, Cleveland, Ohio (September 11-13, 1995).

92. Hans Kruse, "Performance of Common Data Communications Protocols Over Long Delay Links: an Experimental Evaluation," *Proceedings of the 3rd International Telecommunications System Modeling Conference*, Nashville, Tennessee (March 1995).

93.   Shikhar Bajaj, C. Brazdziunas, David Brooks, Daniel Daly, Issac Lopez, Saragur Srinidhi, Thomas Robe, and Faramak Vakil, "Performance of TCP/IP on ATM/SONET ACTS Channel," *Proceedings of the 16th International Communications Satellite Systems Conference*, Washington, D.C. (February 25-29, 1996).

94.   Mark Allman, Chris Hayes, Hans Kruse, and Shawn Ostermann, "TCP Performance Over Satellite Links," *Proceedings of the 5th International Conference on Telecommunications Systems*, Nashville, Tennessee (March 1997).

95.   C. Fair, "TCP Performance Over ACTS," *Transport Protocols for High-Speed Broadband Networks Workshop, Globecom '96*, London, England (November 22, 1996).

96.   Wendy Schmidt, Jeffery Tri, Marvin Mitchell, Steven Levens, Merrill Wondrow, Leslie Huie, Robert Martin, Barry Gilbert, and Bijoy Khandheria, "Optimization of ATM and Legacy LAN for High Speed Satellite Communications," *Transport Protocols for High-Speed Broadband Networks Workshop, Globecom '96*, London, England (November 22, 1996).

97.   Daniel Friedman, Sonjai Gupta, Chuanguo Zhang, and Anthony Ephremides, "Innovative Networking Concepts Tested on the Advanced Communications Technology Satellite," *International Journal of Satellite Communications* 14, no. 3 (May-June 1996) 201-217.

98.   Rodney Long, Michael Gill, and George Thoma, "High Speed Satellite Access to Biomedical Text/Image Databases," *Advanced Digital Libraries Conference*, Washington, D.C. (March 1996).

99.   Deborah Pinck and Loretta Tong, "Satellite-Enhanced Personal Communications Experiments," *International Journal of Satellite Communications* 14, no. 3 (May-June 1996) 249-258.

100.   "Satellite Networks: Architectures, Applications, and Technologies," NASA/CP-1998-208524, *Proceeding of a conference sponsored by NASA Lewis Research Center*, Cleveland, Ohio (June 2-4, 1998).

101.   David Beering, David Brooks, and Michael Zernic, "High Data Rate Experiments," *ACTS Conference 2000 sponsored by the NASA Glenn Research Center*, Cleveland, Ohio (May 31, 2000).

102.   Louis Ippolito, "Propagation Considerations for Low Margin Ka-Band Systems," *Proceedings of the 3rd Ka-Band Utilization Conference*, Sorrento, Italy (September 1997).

103.   David Rogers, Louis Ippolito, and Farimaz Davarian, "System Requirements for Ka-Band Earth-Satellite Propagation Data," *Proceedings of the IEEE* 85, no. 6 (June 1997).

275

104. Dr. Louis Ippolito, "Propagation Effects Handbook for Satellite Systems Design" (prepared for NASA/JPL, 2000).

105. Nasser Golshan, "Interim Findings of ACTS Ka-Band Propagation Campaign," *Proceedings of the 4th Ka-Band Utilization Conference*, Venice, Italy (November 1998).

106. Faramaz Davarian, "Ka-Band Propagation Research Using ACTS," *Proceedings of the 1st Ka-Band Utilization Conference*, Rome, Italy (October 10-12, 1995).

107. Mohamed Alouini, Scott Borgsmiller, and Paul Steffes, "Channel Characterization and Modeling for Ka-Band Very Small Aperture Terminals," *Proceedings of the IEEE*. 85, no. 6 (June 1997).

108. Asoka Dissanayake, "Application of Open-Loop Uplink Power Control in Ka-Band Satellite Links," *Proceedings of the IEEE*. 85, no. 6 (June 1997).

109. Julius Goldhirsh, B. Musiani, Asoka Dissanayake, and Kuan Lin, "Three-Site Space Diversity Experiment at 20 GHz Using ACTS in the Eastern United States," *Proceedings of the IEEE*. 85, no. 6 (June 1997).

110. *ACTS Wide Area Diversity Experiment: Final Report*, (COMSAT Laboratories, SSTD/96-011, March 1996).

111. John Beaver and V. Bringi, "The Application of S-Band Polarimetric Radar Measurements to Ka-Band Attenuation Prediction," *Proceedings of the IEEE*. 85, no. 6 (June 1997).

112. Roberto Acosta, "Wet Antenna Effect on Ka-band Low Margin Systems," *Proceedings of the 4th Ka-Band Utilization Conference*, Venice, Italy (November, 1998).

113. Glenn Feldhake and Lynn Ailes-Sengers, "Comparison of Multiple Rain Attenuation Models with Three Years of Ka-Band Propagation Data Concurrently Taken at Eight Different Locations," *Proceedings of the 3rd Ka-Band Utilization Conference*, Sorrento, Italy (September 1997).

114. Robert Crane and Paul Robinson, "ACTS Propagation Experiment: Rain Rate Distribution Observations and Prediction Model Comparisons," *Proceedings of the IEEE*. 85, no. 6 (June 1997).

115. Robert Crane and Asoka Dissanayake, "ACTS Propagation Experiment: Attenuation Distribution Observations and Prediction Model Comparisons," *Proceedings of the IEEE*. 85, no. 6 (June 1997).

116. Robert Crane, "Evaluation of Global Model and CCIR Models for Estimation of Rain Rate Statistics," *Radio Science* 20, no. 3 (1985) 865-879.

117.   Asoka Dissanayake, Jeremy Allnutt, and F. Haidara, "A Prediction Model that Combines Rain Attenuation and Other Propagation Impairments Along Earth-Satellite Paths," *IEEE Transactions on Antennas and Propagation* 45, no. 10 (October 1997).

118.   Robert K. Crane, *Electromagnetic Wave Propagation Through Rain* (New York: John Wiley & Sons, Inc., 1996).

119.   Roberto Acosta, "Wet Antenna Effects on Ka-Band Systems," *National Radio Science Meeting – URSI 99*, Boulder, Colorado (January 4-8, 1999).

120.   Robert K. Crane, "Four Years of ACTS Propagation Measurements—Comparison With Models," *National Radio Science Meeting – URSI 99*, Boulder, Colorado (January 4-8, 1999).

121.   Thomas Young, "The Best Job in Aerospace," *Issues In NASA Program and Project Management (NASA SP-6101)* 7 (1993).

122.   Norman R. Augustine, *Augustine's Laws* (New York: Penguin Books, 1987).

123.   Dr. Lew Allen, et al, "The Hubble Space Telescope Optical Systems Failure Report," *NASA TM-103443* (November 1990).

124.   *ABC Nightly News with Peter Jennings*, (August 2, 1993). Chris Dixon, senior VP of PaineWebber, and James Pike, head of space policy at the Federation of American Scientists, expressed the view during ABC Nightly News that Ka-band was not needed.

125.   T. Gabriszeski, P. Reiner, J. Rogers, and W. Terbo, "18/30 GHz Fixed Communications System Service Demand Assessment," by Western Union Telegraph Company *NASA Contract Report 159546* (1979).

126.   Robert Gamble, H. Seltzer, K. Speter, and M. Westheimer, "18/30 GHz Fixed Communications System Service Demand Assessment," vol. 2: Main Report by U.S. Telephone and Telegraph Corporation/ITT, *NASA Contract Report 159620* (1979).

127.   Walter Morgan, *Specialized Study of Satellite Program Benefits* (Clarksburg, Maryland: Communications Center, July 1981).

128.   Daniel Kratochvil, et al, "Satellite Provided Fixed Communications Services: A Forecast of Potential Domestic Demand Through the Year 2000, Parts 1, 2 and 3," *NASA CR-168145, NASA CR-168146,* and *NASA CR-168147* (1983).

129.   Daniel Kratochvil, et al, "Satellite Provided Customer Premises Services: A Forecast of Potential Domestic Demand Through the Year 2000, Parts 1, 2 and 3," *NASA CR-168142, NASA CR-168143,* and *NASA CR-168144* (1983).

277

130. Robert Gamble, L. Saporta, and G. Heidenrich, "Customer Premises Services Market Demand Assessment 1980-2000 Part 1 and 2," *NASA CR-168150* and *NASA CR168151*, (1983).

131. Steve Stevenson, William Poley, Jack Lekan, Jack Salzman, "Demand for Satellite Provided Domestic Communications Services to the Year 2000," *NASA TM-86894* (November 1984).

132. Richard Gedney, David Wright, Joseph Balombin, Philip Sohn, W. Cashman, and A. Stern, "Advanced Communications Technology Satellite," *Proceedings of the 41st Congress of the International Astronautical Federation (IAF)*, Dresden, Germany (October 6-12, 1990).

133. Richard Gedney, David Wright, Joseph Balombin, Philip Sohn, William Cashman, Alan Stern, Len Golding, and Larry Palmer, "Operational Uses of ACTS Technology," *Proceedings of the 14th AIAA International Communication Satellite Systems Conference*, Washington, D.C. (March 22-26, 1992).

134. Kent Price and Robert Kwan, "Future Benefits and Applications of Intelligent On-Board Processing to VSAT Services," by Space Systems Loral, *NASA CR 189185* (May 1992).

135. Kent Price, Robert Kwan, Prakash Chitre, T. R. Henderson, and L. W. White, "Applications of Satellite Technology to Broadband ISDN Networks," by Space Systems Loral, *NASA CR 189199* (March 1992).

136. "Potential Market for Advanced Satellite Communications," Booz-Allen & Hamilton, *NASA CR 191145* (June 1993).

137. Grady H. Stevens, "Role of New Technology Satellites in Providing T1 and Higher Rate Services," *NASA Lewis Research Center*, Cleveland, Ohio (January 5, 1994).

138. Richard Gedney, "Satellite Access Techniques for Efficient WWW Communications," *Proceedings of the 4th Ka-Band Utilization Conference*, Venice, Italy (November 2-4, 1998).

139. "Internet-Multimedia Still Growing for Satellites," *Satellite International* 2, no. 2 (January 29, 1999).

140. "Inspired ACTS," *Iridium Today* 4, no. 1 (October 1997).

141. Facsimile from Keith Warble of Motorola to Ronald J. Schertler, ACTS Project Experiments Manager (September 13,1994).

142. Letter from Steven D. Dorfman, President Hughes Telecommunications & Space Sector to Dan Goldin, NASA Administrator (January 25, 1994).

143. Facsimile from Edward J. Fitzpatrick, Vice President of Hughes Space-way to Dr. Richard T. Gedney, NASA ACTS Project Manager (April 25, 1995).

144. Synopsis of the Via Satellite "Satellite 97" Conference, *http://www.phillips.com/ViaOnline/sat97a.htm*

145. Frank Kuznik, "Opening ACTS," *Air & Space* (October/November 1996).

146. Letter from Curtis G. Gray, Vice President, Data and Broadband, WorldCom to Donald J. Campbell, Director of NASA Glenn Research Center (December 15, 1995).

147. Letter from Paul A. Moravek, AT&T InterSpan to Donald J. Campbell, Director of NASA Glenn Research Center (January 10, 1996).

148. Letter from Russell T. McFall, President of Astro Space Commercial, Lockheed Martin to Donald J. Campbell, Director of NASA Glenn Research Center (January 16, 1997).

149. Letter to NASA ACTS Project Office by Dr. John V. Evans, President of COMSAT Laboratories, September 15, 1994.

150. Jeromy Rose and Niall Rudd, "Marketing, Regulatory, Strategic and Technical Aspects of Ka-Band Satellites and Networks," *Proceedings of the 3rd Ka-band Utilization Conference*, Sorrento, Italy (September 1997).

151. Walter Morgan, "Update of Global Ka-Band Filings," *Proceedings of the 3rd Ka-band Utilization Conference*, Sorrento, Italy (September 1997).

152. Joseph Anselmo, "R & D Pipeline Shaping New Era for Satellites," *Aviation Week & Space Technology* (March 31, 1997).

153. Alexander M. Geurtz, "ARCS: A Real World Ka-band System," *Proceedings of the 4th Ka-band Utilization Conference*, Venice, Italy (November 2-4, 1998).

154. Y. Kwon Hwangbo and H. Kim, "The Development of a Tele-Education Service via Koreasat-3," *Proceedings of the 5th Ka-band Utilization Conference*, Taormina, Italy (October 18-20, 1999).

155. W. Baer, L. Johnson, and E. Merrow, "Government-Sponsored Demonstrations of New Technologies," *Science* 196, no. 4293 (1977) 951.

156. Robert Berry, "Power Tools: Responding to the Changing Market for Satellite Manufacturing," *Satellite Communications* 22, no. 4 (April 1998).

157. Ramstad Evan, "Foreign Chip Firms to Boost Funding of Semitech Group," *Wall Street Journal* (1997).

158.   Dean Takahashi, "Chip Firms Face Technological Hurdles that May Curb Growth, Report Suggests," *Wall Street Journal* (December 1, 1997).

159.   Joseph Anselmo and Anthony Velocci, "Wall Street Bulls Chase SATCOM Boom," *Aviation Week and Space Technology* (June 15, 1998).

160.   Walter Morgan, "1977-1997: What a Difference," *Satellite Communications* (October 1997).

161.   Richard McCaffery, "Researchers Fear Lack of Satellite Technology Funds," *Space News* (December 8-14, 1997).

162.   William Broad, "Satellite A White Elephant Some Say," *New York Times* (July 20, 1993).

163.   Bill Clinton and Al Gore, "Technology for America's Economic Growth, A New Direction to Build Economic Strength," (February 22, 1993).

164.   *Satellite Alliance USA: Background and Overview* (American Technology Alliances, September 10, 1997).

165.   Mary Mosquera, "Advisory Group Urges More Federal R&D Funding" *TechWeb* (February 24, 1999).

Web Sites for Information on the ACTS Program

ACTS Description and Results:

http://kronos.grc.nasa.gov/default.asp

ACTS Propagation Program Web Sites:

http://propagation.jpl.nasa.gov
http://weather.ou.edu/~actsrain/

ACTS User/Technology Experiments:
(including propagation plus reports)

http://kronos.grc.nasa.gov/about/experiments/technology/

INDEX